FOOTPRINTS BEYOND EARTH

STEPS BEYOND OUR HOME

HIMANK MITTAL

Made with ♥ on the Notion Press Platform
www.notionpress.com

To the scientists, engineers, and explorers who turned imagination into reality,

and to those who will take the next steps

Contents

Preface

"Remember to look up at the stars and not down at your feet. Try to make sense of what you see and wonder about what makes the universe exist. Be curious."
— Stephen Hawking

Space has always been viewed as distant; physically, financially, and emotionally. For decades, it has been portrayed either as a grand dream or an unnecessary expense. This book is written to challenge both extremes.

Humanity today stands at a transitional moment in space exploration. We possess the technology to reach farther than ever before, yet we rely on infrastructure, systems, and decision-making frameworks built for a different era. Budgets are larger, ambitions are higher, and private participation is growing, but progress often feels slower than expected. This contradiction is not accidental. It is structural.

This book does not glorify space exploration, nor does it dismiss the real concerns surrounding public spending. Instead, it examines space as a system; one driven by money, priorities, risk, and long-term planning. Space exploration is not just about rockets and astronauts; it is about infrastructure, governance, public trust, and the willingness to invest in the future before the benefits become visible.

Rather than focusing on isolated missions, this book follows the flow of capital through space programs and institutions like NASA, while also observing alternative approaches adopted by agencies like ISRO. By studying where money is allocated and where it is not. We can better understand why certain capabilities advance while others stagnate.

It covers the missions we succeeded in, the ones we failed, and the ones in which we can still succeed.

I

I am Himank Mittal, a curious thinker and researcher who loves information. I have worked on four books previously, with a special regard for physics, maths, and social studies. I like to think beyond the norm and write books based on information, facts, and observation.

I enjoy learning about how the world works and questioning ideas that are often taken for granted. Writing helps me organize my thoughts and share what I learn with others. This book has been written by me, based on my understanding and research, and I hope it helps you see space, science, and progress from a broader perspective.

If you like this book, please consider leaving a review on Amazon or Flipkart. Your feedback means a lot and encourages me to keep writing. Thank you for reading.

For any questions or a discussion with me:

Contact me at **mittalhimank736@gmail.com**

PART I

FOUNDATIONS OF SPACE

CHAPTER ONE

Why Space Matters.

What if the greatest threats to humanity do not come from Earth at all?

Asteroids that pass unnoticed, solar storms powerful enough to shut down modern life, and forces we barely understand continue to shape our future from beyond our planet. Space exploration is not driven only by curiosity or adventure; it is a practical effort to understand these dangers and protect the fragile world we depend on.

A widely accepted scientific theory suggests that an asteroid impact caused the extinction of dominant life forms on Earth long before humans existed. Such events are not limited to the past. Even today, asteroids have the potential to threaten human civilization, reminding us that ignoring space does not eliminate the risks it holds.

Many people view these concerns as unnecessary fear, and it is true that pessimism alone achieves nothing. However, space represents more than distant dangers. It offers opportunities that directly affect life on Earth, from agriculture and climate monitoring to communication and disaster management. This chapter explores not only how space already benefits our daily lives, but also what humanity may achieve by advancing further into space.

1. Asteroids and extinction

This is one of the main causes that is brought up when this specific question is raised. Asteroids are very dangerous to all forms of life, whether humans or dinosaurs.

FREQUENCY	IMPACT SIZE	POTENTIAL
EVERY 100 YEARS	20 METERS	SHOCKWAVES
EVERY 1000 YEARS	50 METERS	LOCALIZED DEVASTATION
EVERY 10,000 TO 1,00,000 YEARS	100+ METERS	REGIONAL DEVASTATION
EVERY 500,000 TO 700,000 YEARS	1 KM	GLOBAL DEVASTATION
EVERY 100 Million YEARS	10 KM	MASS EXTINCTION

-i-

Every few years, an asteroid hits Earth and causes a lot of destruction. The most recent one to hit Earth was Chelyabinsk (2013), which landed in Russia and caused injuries to about 1,500 people. In the above table, I have mentioned the size of the asteroid (impact size).

A 10 km asteroid seems small to cause a dangerous mass extinction, but that is only the size of the asteroid. If a 10 km asteroid hits Earth at a speed of 20 km per second, or in SI units 20,000 m/s, or 45,000 mph (which is equivalent to the kinetic energy of billions of Hiroshima and Nagasaki nuclear bombs), it would instantly vaporize a crater of about 180 km wide and 20 km deep.

After that, a literal wall of fire would spread in each direction from the crater, burning everything in its way for thousands of kilometers at the speed of sound (343 m/s).

The "Broiler" Effect

When the asteroid hits, it throws billions of tons of vaporized rock high into the atmosphere. As the debris falls back into the atmosphere, friction turns it into "glass rain" and heats it up. The atmosphere would turn glowing red and become like a pre-heated oven.

The heat radiation would ignite fires in dry vegetation, causing wildfires all over the world. This alone could lead to the extinction of civilization. But if we survive this, the next deadly phase starts.

The Winters

Winters have been a dangerous time of the year for humans for centuries, whether during natural disasters, wars, or prehistoric times. After an asteroid hits Earth, large amounts of dust and smoke would be released into the atmosphere, blocking sunlight for years and causing temperatures to plummet by 10–20°C (50–68°F).

Fire and extreme cooling might sound contradictory, but it actually works like a seesaw. In the first two hours of the asteroid impact, the broiler effect would destroy most of the planet. After that, a rapid cooling effect would impact Earth. A huge cold wave would affect Earth for about a month.

After the cold wave, the temperature would stabilize into a deep freeze as sunlight remains blocked. This could last for around 10 years. During this time, plants and vegetation would die, and global starvation would be a real possibility.

After 10 years, the smoke and dust would slowly fade, and greenhouse gases would gradually re-warm the planet. Oceans play a major role in this process, as they stay cooler for longer while land heats up faster.

What have we done?

Now, a question arises: what have we done to avoid such a mass extinction?

Step 1: The Map

We cannot stop what we cannot see. NASA and its international partners have identified 95% of asteroids larger than 1 km that are capable of causing mass extinction.

Good news: None of them are on a collision course with Earth for the next 100+ years.

NEOWISE is a space telescope used to identify dark asteroids, which are usually difficult to detect.

Step 2: The Avoidance Technology

In September 2022, NASA tested DART (Double Asteroid Redirection Test), which was the first-ever interplanetary defense mission.

What happened: NASA crashed a spacecraft the size of an arcade machine into a small asteroid called Dimorphos at 22,500 km/h.

The result: It worked better than expected. It shortened the orbit of the asteroid by 32 minutes. This proved that we do not need nuclear weapons to save Earth. If we spot an asteroid early enough, we can hit it with a kinetic impactor to deflect it and nudge its path.

Step 3: Early Warning System

The asteroids we know are not the real threat; the ones we do not know are the real dangers.

NEO (Near-Earth Object) Surveyor, launching in 2027, is an **infrared** telescope that would specifically detect asteroids 140 m or larger, hidden by the Sun's glare.

Step 4: Emergency Options

If we find an asteroid too late to use a deflection option, scientists have two main backup plans:

1.Gravity Tractor:

A heavy spacecraft flies near the asteroid, and its small gravitational pull slowly moves it away from its path toward Earth. Although it is very slow, it is very precise.

2.Ion Beam Shepherd:

This method uses a beam of ions to push the asteroid away over a long period of time.

Conclusion

All of this combined—observation, action, and knowledge—which is all part of our space investments, provides a future-proof plan for such disasters in the future.

2. Rare Earth Elements (REE)

"The first trillionaire there will ever be is the person who exploits the natural resources on asteroids."
— Neil deGrasse Tyson

Rare Earth Elements are materials that are not present in very high quantities on Earth. We already have a shortage of these materials on our planet. However, outside Earth, these elements are present in huge quantities on asteroids, and some celestial bodies are even made up of them.

Space exploration has the power to bridge the gap between us and these precious minerals. If we are able to extract these resources, they would last for centuries, and resource shortages would not be a major problem.

Sadly, we are not able to extract the full potential of this discovery because the economics do not make sense yet. The travel and return of those materials cost too much. This part is explained briefly in the next chapter.

If this were a reality, we would already have many trillionaires. Every asteroid has elements worth billions, so even mining a single asteroid could make an individual extremely rich. Today, space research focuses on finding cheaper alternatives for space transport so that mining becomes viable.

Can anyone own an asteroid?

The answer is no. However, space laws allow companies to keep the money earned by mining asteroids. So, companies can technically profit from asteroid mining.

A potential REE extraction process

First, an electric rocket could be very viable for the project because it does not need much fuel once in space. Although it is harder to launch, it becomes very efficient in orbit.

Next, we would target Near-Earth Objects (NEOs). Although we have reached asteroids far away from Earth, capturing one near Earth for the first mission is a smarter choice.

To capture the asteroid, the satellite could use multiple options, such as vaporizing the rock or using thrusters. Then, the asteroid would be brought onto the right trajectory toward the Moon. Using the Moon's gravitational pull, we could secure an orbit around Earth.

Meanwhile, an Extraterrestrial Resource Extractor (ERE) would be launched into the same orbit. Once the ERE meets the asteroid, it would create a protective shield around the mining area and use mirrors to vaporize the rock into gases. Drills would also be used. Centrifuges would separate light elements from heavy ones.

Finally, a simple capsule could be launched toward Earth, which would land in water and be collected by ships.

What do we really have on other planets?

1. **Mercury (The Iron King):**Mercury has a massive **iron-rich core** that occupies about **57%** of the planet's mass. It also has nickel and cobalt in high amounts.
2. **16 Psyche:**Although an asteroid, this metallic body between Mars and Jupiter contains **gold**, iron, and nickel worth about **700 quintillion dollars**, enough to make everyone on Earth a billionaire.
3. **Titan:**Saturn's moon Titan has **30 times more gas reserves** than Earth and contains massive lakes of methane and ethane.
4. **Europa and Ganymede:**These moons of **Jupiter** have massive subsurface oceans with more liquid water than **all the oceans** on Earth combined.
5. **55 Cancri e (The Diamond Planet):**This planet outside our solar system is considered a "super-Earth" and is very carbon-rich. Due to extreme pressure, it is believed to be largely made of crystalline diamonds.
6. **Luna (Our Moon):**The Moon contains **Helium-3, a rare isotope** that could help us build clean nuclear reactors in the future.
7. **Jupiter and Saturn:**These planets are massive reservoirs of **hydrogen and helium**, although extracting them from their intense gravity remains a major challenge.

This data shows how rich our solar system is in terms of resources, and this does not even include asteroids. If we learn how to use these resources properly, we could advance further as a civilization. Mining in space would reduce damage to Earth and help protect our home planet.

Although the logistics are extremely difficult, solutions are discussed in the next chapter.

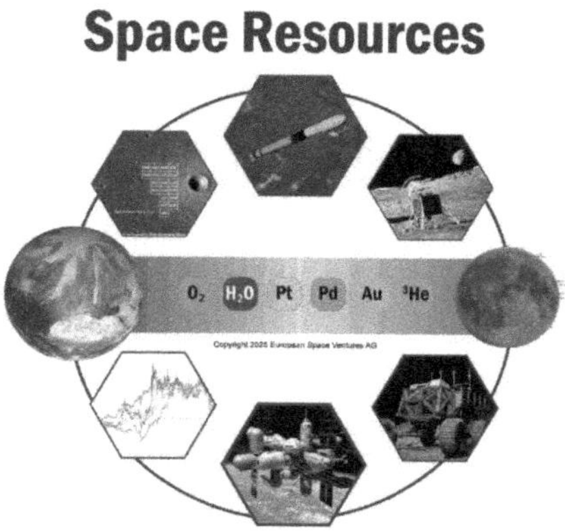

-ii-

3. Space Innovations in Everyday Use

Communication satellites are major contributors to our daily lives. Most rural areas or regions where cable connectivity is impractical are now covered by satellites.

SpaceX has a massive satellite network. About 10,000 of the 15,000 satellites in space are owned by SpaceX. These satellites allow navies and airplanes to maintain communication, and they make internet connectivity in villages viable.

Weather Satelites help us to see the atmospheric conditions on earth and pre-inform people for disasters, storms, and conditions that could effect their daily routine. This is a very complex yet helpful tech that help us in our day to day lives.

Observation Satellites allow us to observe unusual patterns or activities on our planet. Whether Stubble burning(Parali) or any other activity that causes disturbance in our ecosystem, these satellites provide alot of data on the patterns we are unable to observe from our eyes. These satellites have brought Human Research and Development even further.

CHAPTER TWO

ECONOMICS OF SPACE

Today is a very critical time for human investment in space. The ISS (International Space Station) has almost reached the end of its operational life. People are questioning investments in space, and there is a severe crisis of space suits for new missions. Private companies are investing more than ever, and the idea of mining in space is slowly dying. Money spent on space technology is at an all-time high, yet the development is not noticeable. The money is flowing, but the results are not visible. We are seeing a transition phase from old systems, but the newer versions are not ready. A governance shift is taking place, from public to private. In this chapter, I explain the flow of money in space.

Part 1:The Transition Phase

Sub Section-1:Public Skepticism; Funding Highs But Visible Stagnation-

- In the humble beginnings of human space exploration, missions were mainly supported by governments. Today, the situation has drastically changed. This is largely because of the huge satellite broadband investments made by companies like SpaceX.
- However, even today, governments still spend a major portion of global funds on space technology, as it is recognized as the future of human settlement and advancement of civilization. NASA contributes a significant part of global space spending. In FY2025, NASA was allocated $24.4 billion, while global space spending reached an all-time high of

$137.4 billion.

- Out of this $137.4 billion, 54% is allocated toward defence-related space activities. This indicates that space is still viewed more as a strategic battlefield rather than purely as a settlement option. A balanced perspective is essential to understand the full picture.
- For FY2026, the White House suggested a 24% budget cut for NASA. However, Congress reduced this to a 1.88% decrease from FY2025 levels. While this is not a drastic cut, it may reflect shifting priorities.
- The United States, a major player in space, showing even slight hesitation in investment raises questions about the future pace of advancement. Meanwhile, other nations are increasing their efforts to gain advantage in this evolving space race.
- Russia, like the US, is not pushing on space enough. Russia plans to spend a sum of $4 billion on this sector. It is still the fourth highest globally, but Russia, as a historically major contributor to space development during the Space Race, could use this legacy to its advantage for further progress in space.
- It has been almost 100 years since humans started investing in space, yet many people have still not realized that it is not a con, but a pro.
- This gap between the thinking of the government and the people is leading to a significant slowdown in new technologies. The government has to think twice about whether an expense for a space mission should be presented in the budget or not, and whether people would agree with it. The whole space industry is suffering from a "funding crisis."
- If we want to reach Mars by 2050, we would need cooperation. Otherwise, such funding-intensive missions are not financially feasible with the current level of support from the government.

Sub-Section-2:The Infrastructure-

- Due to budget shortages, the agencies have been using much of the same infrastructure since the 1960s. This has led to a range of operational problems. The equipment has degraded over the years and needs to be replaced urgently. We are in a turbulent transition phase, where the newer machinery is not yet ready, and the older systems are gradually failing.

- The ISS has almost reached the end of its operational life. It needs to be replaced, as it acts as a powerful hub for humans in space. After Earth, the safest place in the universe for humans to access is the ISS. If any astronaut is stranded in space, the first option for rescue is to help them reach the ISS. It acts as a docking station and a refueling hub for larger missions that require massive amounts of fuel.
- It acts as a hub for deep scientific research and development, and for conducting experiments. If this is not resolved, it would halt progress and cause a huge loss to the research and development industry. Older equipment is maintenance-heavy, requiring frequent fixes, which drains a lot of money that could have been used for better purposes. Already operating under a tight budget, optimization becomes the key.
- If we truly want to spend our money correctly, we need to renew infrastructure up to modern standards, equipped with newer technology, which would further accelerate growth. A non-functional ISS would offer less operational flexibility and make us more risk-averse. Mission planning would become conservative, and we would lose one important safeguard in space for our "delicate" rockets. Planning and testing of rockets would become much more rigorous, draining additional funds, and the success rates of such missions would drop drastically.
- A non-functional ISS just floating in low Earth orbit could also become a threat, both to Earth and to other objects in orbit. Once the ISS leaves the spectrum of human control, it could become a threat to the thousands of satellites in low Earth orbit and potentially cause serious damage.
- The daily astronaut expeditions we send to space so easily today have been made possible by the convenience of the ISS. In just a few years in space, it has created many opportunities for discovery and has helped countless astronauts reach orbit. It has made living in space for six months much easier.
- Regular launches to the ISS have tested our launch and docking capabilities multiple times. The renewal of the ISS is not just an option, but a question; a question that asks whether we want to quit space or not.
- If you and I can clearly state the problem, the effects, and the solution, then why can't it be implemented?
- Well, the problem in the solution lies in capital. As I have previously stated, the expenditure for these projects is very high, and to receive this money, the agency needs to make a request to the government. The government cannot allocate such a large amount at once, as it would

disturb the entire budget cycle. Money would have to be cut from many sectors, especially those where most public spending is directed, such as social welfare, defence, education, and health.

- If such a proposal is presented to the general public, the government would face major public outrage. As we know, democratic governments often act in line with public opinion rather than what may be purely beneficial in the long term. These projects require large sums of money and long timeframes, which agencies are often not ready or able to commit to at once. At present, much of the attention is focused on large human missions, such as human settlement on Mars and bringing humans back to the Moon.
- Higher maintenance costs are sometimes seen as more acceptable for agencies, as smaller annual expenses appear more manageable than a single, large capital expenditure.
- In the vastness of space, the primary factor that makes it uninhabitable is the scarcity of oxygen. Most living organisms on Earth require oxygen to sustain their day-to-day lives. If space contained Earth-like air everywhere instead of a vacuum, it would be much easier to settle in space. To solve this problem and make space travel possible for humans, we invented a revolutionary solution — the space suit.
- The space suit has enabled significant advancements in space exploration and science over just 60+ years and has helped us become a far more advanced civilization.
- But these essentials for space travel have become really expensive and unaffordable in the current inflated market. Government agencies are also facing pressure from rising prices of essential components. Every part of a space suit has to be of extremely high quality and well above standard requirements. These suits undergo a series of tests, as the lives of astronauts literally depend on a single suit.
- A space suit contains many advanced and out-of-the-box components and extraordinary innovations, through which it is ensured that, under any possible condition an astronaut might face, they do not experience discomfort. The personalization and explanation of every single component in a space suit is discussed in detail in Part III, Chapter 6: Human Life in Space.
- The next generation of space suits must be advanced enough to support another 50 to 70 years of human progress and future missions. The future of space is increasingly directed toward human settlement, which

would require the mass production of such suits. Space tourism is also bound to begin — not today, but someday, and that day is approaching. For long-duration missions, comfort in space suits becomes essential. After all, an astronaut is still a human being and needs interaction and movement.

- The space suits we use today were designed for short journeys to space, not for long-term settlement missions such as missions to Mars.
- Adding all these features, along with larger oxygen tanks and better waste disposal systems, would require a lot of spending allocated to research and development. Once the technology is developed, it would need to pass multiple tests on Earth's surface, then in the sky, in oceans, and finally be tested in space. It would also require approval from multiple regulatory bodies, and implementation and adaptation are another long process altogether.
- This entire procedure, from development to prototyping, testing, regulatory approval, setting up manufacturing, and finally training and adapting astronauts to the new systems - would take more than two decades and involve massive, often unbearable expenditures.
- However, this is not even the principal obstacle. The root cause of this issue is the age of the existing space suits. We have been using the same suits throughout the years. Just imagine wearing the same clothes for 50 years. This is the present-day reality of space. Astronauts have to use the same suits that someone else wore 50 years ago.
- NASA is not able to manufacture new suits and provide the level of comfort astronauts need today. Life today is very different from then, and the requirements are much broader. Space suits are extremely high-maintenance, even from the day they are manufactured. Now imagine the cost of maintaining such old suits within an already limited budget.
- A single malfunction in a suit could mean the instant loss of life. It is the most fragile layer of defence in space for an astronaut. An astronaut literally carries their life on themselves. A settlement on Mars or the Moon means that a space suit would no longer be temporary equipment, but daily infrastructure. It should not be too tight or uncomfortable. It should be reliable and easy to use, to the extent that even an untrained person could theoretically operate it.
- Replacing these older suits would mark the beginning of a new era of space advancement, a time in which the life of an astronaut would no longer mean suffering, but a more stable and sustainable experience. It

would be a time in which we could think of farther missions and no longer face such fundamental obstacles in space travel. A time where we could think larger, without constant limitations. A new generation of space suits would open doors to multiple opportunities that we are currently restricted from exploring.

- These are all major space-based infrastructures that need renewal. However, the least attention is often given to smaller issues that actually matter a great deal.
- Today, the ground reality is even harsher. The base infrastructure remains largely the same, but the demands have increased. Yes, upgrades have been made, but they are not sufficient to sustain the rapid growth path we are moving toward.
- Launch services are in particularly poor condition. Many major launch pad complexes date back to the 1960s–1980s. Aging fuel lines, old flame trenches, legacy control rooms, and maintenance-heavy structures are still in use. These systems do not offer the reliability and control required today and create a constant fear of disaster. Aging fuel lines are still being used, which could result in fuel loss. Control rooms remain largely unchanged; although they are equipped with some new systems, many older components are still in operation. The same launch pads have been used for thousands of launches, including different types of missions such as ISS re-equipment, test missions, lunar missions, satellite launches, and even rescue missions. Such aging systems require regular maintenance, often consuming a large portion of an already limited budget. Modern rockets may be new, but the supporting systems are old.
- Another major problem lies not in hardware, but in software. The same legacy systems are still used across multiple missions, requiring far more professional handling than modern, simplified systems would. Security is also not fully up to date, leaving many missions vulnerable to malware attacks. Data is still transmitted through older communication infrastructure, including satellite ground stations and telemetry, tracking, and command (TT&C) systems. These systems, which act as the communication backbone for satellites and rovers, have aged significantly and often result in connection outages.
- The next problem arises in testing equipment. Vacuum chambers, vibration tables, and thermal testing labs are decades old and are in urgent need of innovation. Everywhere, we see people suggesting technologies for missions to Mars or the Moon. However, we often

neglect the smaller innovations required at the ground level.

- Every small innovation compounds over time and often leads to larger, more efficient breakthroughs. Many major advancements would not function as effectively or remain future-proof if these smaller improvements were ignored. In the book Atomic Habits, written by James Clear, the example of the British cycling team is discussed, showing how compounding small habits led to massive victories. Even a 1% improvement made every day over a year can result in a 37.5% overall improvement.
- In space, every single percentage matters. Such advantages could save large amounts of fuel and make space missions easier and more worry-free to operate. The risk to an astronaut's life would be reduced, and agencies would not have to hesitate before undertaking missions involving human lives.
- The manufacturing plants for these rockets are old, equipped with outdated tooling and multiple manual-intensive procedures. They have lower capacities and operate with higher risks. These older plants take long periods to manufacture rockets. Right now, we are not launching rockets daily, but when that time comes, this will become a major obstacle.
- We need to be fast enough to manufacture a rocket within 10 days. Just imagine an asteroid heading toward Earth, expected to arrive in two months, while a single rocket takes six months to manufacture. Speed and production capacity are the real issues.
- Older plants are still unable to fully integrate the new innovations that our engineers contribute every day, as replacing the entire system is often necessary rather than simply upgrading parts of it. Each upgrade weakens the foundation on which it was built. A completely new plant with world-class facilities must be constructed so that we can prepare rockets at scale.
- All of this chaos is part of the first transition phase. Money shortages and a lack of attention are the primary reasons for this turbulent time. We need to widen the scope and look beyond the present situation. Changing existing infrastructure with better alternatives will be dramatic, but it is bound to happen.
- This is not an issue; it is a solution forcing itself forward. I am only creating a sense of urgency, because it is needed. We need solutions, not panic. Let us stand together for the issues that truly matter - not for the

success of any single person or company, but for humanity as a whole.

-ii-

Part 2: Fund Allocation

In this part of the book, I cover where the money actually goes and what it truly means for us and for the development of civilization in space. I will present how much money flows into each sector, along with a complete breakdown and summary.

On the basis of this information, the next sections will be expanded, and you can form your own assumptions while considering my opinions.

The budget breakdowns are arranged in descending order of budget value. Only three Agencies have been included, who I believe shall be valuable to explore and cover and to not waste much of your time. Budget sources for other agencies have been included in the bibliography, If you are interested, you can read about it more.

BUDGETS

NASA

NASA's budget this year was **$24.4 billion(0.3% OF US GDP),** which is not very large, but not small enough to ignore. It, As mentioned previously, this figure shows a major discrepancy from the budget suggested by the White

House. In fact, it represents the biggest discrepancy in U.S. history since the 1978 budget, which had a 30% discrepancy. This was a massive win for NASA in such a tense situation.

This discrepancy was primarily caused by the major government shutdown beginning on October 1st and lasting 43 days, with the budget being signed by Trump on November 12th. Multiple compromises were made, and NASA managed to make minimal sacrifices.

These numbers refer to the Presidential Budget Request, not the final appropriated budget passed by Congress. These numbers show intent, although the real numbers are similar, with a higher bdugt allocated towards artemis.

The total budget proposed was **$18.8091 million**, which opened doors to multiple new projects and missions that NASA is considering or currently working on.

Exploration

The first focus is naturally directed toward exploration. The foundation of space investment lies in the idea of exploration — discovering worlds that were previously unknown to us. One of the main reasons people support spending on space is exploration, making it important to allocate a significant portion of the total budget toward it.

Missions to Mars are currently in the spotlight, and NASA has kept this under serious consideration. For Moon-to-Mars transportation systems, **a $4.8946 billion** budget was allocated. This ensures that efforts toward Mars exploration continue. It also reflects NASA's approach to Mars - not directly from Earth, but through the Moon. This indicates an intent to expand existing capabilities rather than starting entirely from scratch, as seen with SpaceX.

The **$4.8 billion** allocation was primarily for reaching Mars. However, multiple additional systems must be developed to sustain life there. Therefore, **$2.8154 billion** was allocated to the Moon-to-Mars Systems Development budget, enabling complete development of the entire process rather than just individual components.

Human Exploration Requirements and Architecture received further support of **$602.9 million**, ensuring that the necessary needs for exploration are met. In total, **$8,312.9 million** was proposed for space exploration as a whole, supporting long-term development goals. A strong emphasis was placed on Mars exploration, indicating that it is not merely a symbolic effort, but a financially backed mission focused on research and development.

BUDGET-NASA

LIST(NUMBERS IN MILLION $)	FY25	FY26	FY27
MOON TO MARS TRANSPORTATION SYSTEMS	4,213.0	4,894.6	4,697.7
MOON TO MARS SYSTEMS DEVELOPMENT	3,288.1	2,815.4	2,864.2
HUMAN EXPLORATION REQUIREMENTS AND ARCHITECTURE	117.5	602.9	751

-iii-CHART 1

Space Operations

Next, NASA has allocated funding toward operations in space. Maintaining existing operations is essential. Practicing human presence in space daily strengthens our systems and builds trust in our launch capabilities and rockets. Sustaining the ISS helps prepare astronauts for the future and normalizing living in space, so an initial number of people can eventually be settled on Mars. This year, **3.1319 Billion USD** was proposed for sustaining existing space operations.

Providing equipment, food, and sustaining ISS operations is a mandatory cost every year. This year, **920 Million USD** was proposed for sustaining ISS operations. Experiments are performed on the ISS daily and therefore it is necessary to maintain them.

1.2938 Billion USD was proposed toward space transportation. This includes private flight services purchased from companies like Boeing and SpaceX to transport astronauts to the ISS. A major part of this is to be allocated toward building the SLS rocket and Orion spacecraft meant for the Artemis mission mentioned in **Part 3, Chapter 4: What Have We**

Accomplished?

To support space and flight operations, **645.8 Million USD** was proposed.

As government agencies are facing budget shortages, they are trying to maximize private collaboration rather than taking full financial risk themselves. **272.3 Million USD** was proposed for Commercial LEO development, including satellite infrastructure and spaceflight missions. Although a small amount, it indicates NASA's continued interest in private collaboration.

The total budget allocated for space operations turns out to be **3.1319 Billion USD**. This is to make sure human presence in space does not end, as once it does, like after the Moon landings, it becomes difficult to restart operations and requires significant capital expenditure and willpower.

BUDGET-NASA

LIST(NUMBERS IN MILLION $)	FY25	FY26	FY27
ISS	1,269.6	920.1	920.1
SPACE TRANSPORTATION	1,862.1	1,293.8	1,263.8
SPACE AND FLIGHT SUPPORT	1,088.4	645.8	645.8
COMMERCIAL LEO DEVELOPEMENT	169.6	272.3	302.3

-iv-CHART-2

Categories without breakdown

For space technology, NASA proposed a **$568** million budget, a significant decrease from last year, indicating a shift in focus toward major core missions rather than newer initiatives. Aeronautics, a field where NASA works on flight-related technologies, was allocated **$588.7** million. This is a relatively different domain from NASA's primary space-focused missions, although NASA has contributed many successful innovations in this area.

In aeronautics, NASA primarily works on engine technologies, developing faster, more efficient, and electric propulsion systems, while also providing support to an already flourishing aviation industry.

This year, NASA proposed a **$0** budget for STEM engagement -yes, **zero**. This decision is part of NASA's cost-cutting measures, though it is likely to disappoint many students. It is a risky decision that could cause significant backlash, as it signals a complete withdrawal from NASA's education programs.

The Office of the Inspector General was allocated **$40.7 million**. This office is responsible for identifying technical difficulties, risks, and operational issues before and during missions, helping prevent major failures before corrective action is required.

BUDGET-NASA

LIST(NUMBERS IN MILLION $)	FY25	FY26	FY27
SPACE TECHNOLOGY	1,181.8	568.9	568.9
AERONAUTICS	935	588.7	588.7
STEM ENGAGEMENT	143.5	0	0
INSPECTOR GENERAL	47.6	40.7	40.7

-v- CHART-3

Science

Apart from focusing on space missions, NASA is involved in a wide range of experiments across different scientific fields, including those conducted in space. These experiments help us understand the behavior of various objects, animals, and physical properties, and how they react in space. These sciences allow us to study multiple phenomena and understand how space influences Earth.

This year, NASA proposed a **$3,907.6 million** budget for science. Out of this, **$1,035.9 million** was allocated to Earth Science, supporting experiments conducted on Earth or for understanding Earth itself.

Planetary science was allocated **$1,891.3 million**, aimed at exploring and understanding the behavior of planets and gathering detailed information about them. This knowledge will directly support future missions to Mars and beyond. Only by properly understanding other planets can we realistically plan long-term human settlement.

Astrophysics is an essential field for the advancement of space exploration. Without progress in astrophysics, we cannot design rockets capable of efficiently traveling across the solar system or establish sustainable bases on Mars. For this purpose, **$523.0 million** was allocated.

Heliophysics focuses on the study of the Sun. The Sun plays an essential role in sustaining life on Earth, and in a world increasingly dependent on satellites, monitoring solar activity becomes critical. Solar storms and other solar phenomena can strongly influence Earth's atmosphere and satellite infrastructure. Therefore, **$432.5 million** was proposed for heliophysics research.

Biological and physical sciences are also essential for understanding how the human body reacts in space and how astronauts and future settlers may be affected. This research helps prepare for long-duration missions and supports the development of improved space suits. For this purpose, **$25 million** was allocated.

BUDGET-NASA

LIST(NUMBERS IN MILLION $)	FY25	FY26	FY27
EARTH SCIENCE	2,378.7	1,035.9	1,055.9
PLANETARY SCIENCE	2,731.5	1,891.3	1,861.3
ASTROPHYSICS	1,578.1	523.0	543.0
HELIOPHYSICS	786.7	432.5	422.5
BIOLOGICAL SCIENCES	90.8	25	25

-vi- CHART-4

Safety, Security and Mission Services

Safety, security, and mission services are also mandatory expenses. A total of **$2,118.3 million** was proposed to be allocated for these areas. This budget is further broken down into two sections:

Mission Services and Capabilities - Mission services include systems that guide rockets after launch, manage satellites in space, and control other missions. Capabilities refer to the long-term infrastructure built to support these activities. **$1,498.0 million** was allocated for this purpose. If this amount is not spent, a lot of breakdowns could occur, causing many disruptions in the industry and chaos all over.

Engineering, Safety, and Operations- Building systems, maintaining safety, and keeping operations running also come at a cost. **$620.3 million** was allocated to this field as well. Salaries of employees have to be paid from it and safety today is not a luxury, but a norm.

BUDGET-NASA

LIST[NUMBERS IN MILLION $]	FY25	FY26	FY27
MISSION SERVICES AND CAPABILITIES	**2,058.1**	1,498.0	1,671.8
ENGINEERING, SAFETY AND OPERATIONS	986.3	620.3	446.5

-vii- CHART-5

Construction and Environmental Compliance & Restoration

Construction and environment were granted **$140.1 million**. Construction is a continuous responsibility rather than a temporary task. It includes the development of new facilities and the upgrading of infrastructure. The more modern a facility is, the more it can contribute — both to development and to its employees. **$110 million** was allocated toward construction.

Multiple facilities also get damaged during rocket launches, for which restoration is required. Provisions for such repairs are included in the budget.

Recently, environmental awareness has become a prominent issue, and large organizations are increasingly required to comply with new regulations. To follow these guidelines, NASA has allocated **$30.1 million** for environmental protection.

BUDGET-NASA

LIST(NUMBERS IN MILLION $)	FY25	FY26	FY27
Construction	346.2	110	105
Environmental Compliance	79.4	30.1	35.1

-viii- CHART-6

This year's budget for NASA placed a strong emphasis on missions to Mars, existing operations, and development in the sciences, defining NASA's path forward. The agency appears increasingly focused on the exploration of outer space, with funding allocated toward the development of multiple rocket systems and survival systems.

Notably, NASA does not appear to be prioritizing the development of a new space station, potentially signalling that the next space station may not come from NASA, but from other agencies such as ISRO, which seem more interested in that area.

CNSA(Chinese Space Agency)

The Chinese government is considered to be the second-highest spender in space. It is estimated that **$20.6 billion** was allocated for this year's budget. This is approx **0.11%** of the chinese GDP. These numbers may differ from other sources, as the data for the Chinese space program is based on my own research and analysis, along with estimates from sources such as Statista and others, rather than official figures released by the Chinese government.

The government of China has invested a lot in space technology and is therefore receiving many positive developments as well. Although the agency started later in history and was not part of the original space race, it has caught up with the leading spacefaring countries within just a few years. This is an example of how the right ambition and leadership can help a country develop significantly, even when it enters the race late.

While NASA and other agencies appear hesitant in certain areas, China continues to develop and close the gap. It has developed multiple new technologies and rockets, which have helped it grow further. These efforts have also supported China's self-reliance and overall national growth.

Out of the **$20.6 billion budget**, approximately **$12.8 billion** is allocated to civilian space and exploration activities, which include general space missions. The remaining **$7.8 billion** is allocated to military and national security space programs, for which limited information is publicly available.

Because the Chinese government does not release its data publicly, these numbers are well researched but not 100% accurate, although they are close to the real figures.

The major investments from the **$12.8 billion** civilian space budget are as follows:

1. Human Spaceflight-

This is the largest share of China's space spending. It amounts to approximately **$4.5–5 billion.** It includes the **Chinese Manned Space Agency (CMSA)** and the **Tiangong Space Station**. This is a good amount of spending for space and clearly alloted towards specific missions, rather than scattered all over the place. This helps them to target on specific missions and get them completed, without any problems.

1.Lunar Hardware-

This includes the development of the **Mengzhou crewed spacecraft** and the **Lanyue lunar lander**, both aimed at achieving China's 2030 Moon landing goal.

2.Station Operations-

This covers the maintenance of the three-module T-shaped space station configuration, including cargo resupply missions **(Tianzhou)** and crewed missions **(Shenzhou)**.

2. Lunar and Deep Space Exploration-

The China National Space Administration (CNSA) manages the robotic Tianwen and Chang'e programs, which together cost approximately **$3–3.5 billion**. These missions help us explore deeper into space and understand its properties, along with many aspects that are still unknown to us.

Chang'e 7/8 — These missions plan to land near the lunar south pole by 2026/2028 and establish a research station there.This research station wil help us to collect much more information about the moon.

Tianwen 2/3 — These missions focus on asteroid sampling and a Mars sample return, targeted for 2028/2030.

3. Space Infrastructure and Science-

For space infrastructure and science, an estimated $2.5–3 billion has been allocated toward infrastructure development and scientific programs.

Xuntian Telescope — The Chinese Survey Space Telescope (CSST), a flagship optical observatory scheduled for launch near the space station.

Gaofen Program — Continued deployment of the high-resolution Earth-observation constellation.

Beidou Maintenance — Sustaining the Beidou Navigation Satellite System (BDS) for civilian use.

4. Commercial and Industrial Use(1-1.5 Billion)

National Funds — Establishment of a national commercial space development fund to integrate private entities into national missions.

Infrastructure — Building new commercial launch pads and opening space debris data to the private sector.

Conclusion-

China is more mission-focused and has invested heavily in fields that matter. It has allocated funds for private entities, which is a positive sign, and is establishing multiple stations and observatories, creating several hubs of research in space instead of relying on just a single permanent one.

China is also involved in building its next space station and has invested heavily in it. As it operates independently, it can focus on missions it considers important.

ISRO(Indian Space Agency)

Indian Space Research Organisation's budget for FY 2026–27 is approximately **$1.65 billion**. India is still far behind in terms of space spending, which becomes a major disadvantage compared to other countries. Due to this limitation, ISRO is sometimes unable to attract top-tier talent, as higher compensation expectations may not always be feasible within its budget constraints. Missions are therefore sometimes compromised in scale or pace, which can create operational challenges later.

Yet, ISRO has found ways to operate on lower budgets while achieving a significant percentage of what larger agencies have accomplished.

Its allocation for central schemes and projects is **$1.546 billion**, forming the major share of its total budget. ISRO has managed to reduce expenditures in other areas and focus on what it considers core priorities.

Space technology was allocated **$1.25 billion** from these central schemes. This is the core engine of ISRO, from which most major missions are supported.

This includes:

-Launch Vehicles

LVM3 (Gaganyaan class)

SSLV (Small Satellite Launch Vehicle)

-Satellite Development

-Human Spaceflight (Gaganyaan)

Crew Module

Life Support Systems

Crew Escape and Safety Tests

-Infrastructure

Launch Pads (Sriharikota)

Tracking, Telemetry, and Control Networks

Integration and Testing Facilities

India faces multiple national challenges, including security concerns, extreme weather conditions, droughts, and regional tensions. Around 70% of the population is still directly or indirectly dependent on agriculture. Space applications were allocated $208 million to address these challenges by deploying satellite networks and strengthening India's self-reliance. This includes support for NavIC, satellite communication, disaster management, and remote sensing technologies that assist citizens in their daily lives.

Space science was allocated $68.7 million, while $15.8 million was provided for operating the NSAT and GSAT networks.

ISRO, like CNSA, is more mission-focused than NASA. NASA remains in the global spotlight and plays a major role in maintaining the International Space Station and conducting various scientific experiments in space, which influences its spending priorities.

I have not included other major space agencies such as Roscosmos for multiple reasons, mainly to keep the book focused and engaging while still providing relevant information. Due to the ongoing war in Europe, several countries have diverted portions of their budgets toward defense and related priorities. Because of this shift, I do not consider their current space spending to be as stable or representative for long-term comparison, and including it would only add length without adding meaningful value to the analysis.

Summary:

While NASA's budget is all-inclusive and covers everything from science and missions to operating existing infrastructure, India and China do not have the burden of managing such large and complex operations. ISRO is better focused on the well-being of the entire nation and on delivering direct impact to citizens, while also working on missions such as Gaganyaan and the Bharatiya Space Station.

CNSA, on the other hand, is more focused on developing multiple centres and checkpoints in space, which could later help build a sustained base for China across future missions. This is something NASA was unable to maintain after Apollo, leading to the loss of its continuous presence on the Moon.

Private Funding In Space

Governments have historically been inefficient when trying to push forward sectors managed entirely by themselves. For the past 60 years, many governments carried the weight of the space sector alone. This created a growing gap between the public and the government; between funding and visible results. When that gap widens, the private sector enters the field.

Private companies bring hope, reassurance, and the promise of greater transparency, capital, efficiency, and innovation. However, whether this is fully achieved depends on the institution itself. When the private sector enters an industry, a phase of rapid development often begins. One person can take one step at a time, but thousands of people working together accelerate progress.

In the last ten years, a lot of space startups have opened, and governments are now more than ready to collaborate rather than take the whole risk on themselves. Major startups have emerged, especially in India and China, where they are being given multiple opportunities in developing economies. Although SpaceX has dominated this field, mainly due to the massive funds available to it from the US government and investors, other startups have also developed technologies that we could not achieve in so many years.

In space, the entry of private companies is a large positive implication for economics, as they are profit-driven and therefore can better figure out the economics of space and missions like resource extraction from asteroids. Last year, SpaceX earned 16 billion USD, out of which 8 billion was profit. Large companies are efficient and work continuously throughout the year, unlike governments, and are less likely to make mistakes that lead to loss of money, which they do not prefer.

Private companies are the future of space development, efficiency, and systems, and we must support them. Sure, there are disadvantages, but they will certainly support many missions that are not possible today simply because of funding limitations.

Part 3: The History

History often seeks the truth. To make assumptions about the future, the past is a crucial factor that speaks more clearly than baseless assumptions.

Therefore, viewing the cost factors of previous missions is a necessity to understand the next section and gain a better perspective of how missions actually work.

NASA has spent a total of $650 billion historically, a significant amount toward the expansion of human activity in space. A major portion of NASA's historical spending was allocated to its largest project, the Apollo Program. These missions were a major success for the geopolitical landscape and for human progress in space.

Approximately $280 billion (inflation-adjusted) was spent on this historic achievement. However, each mission cost an estimated $3 billion. This reflects a more difficult reality. Only a small number of space agencies today have annual budgets higher than $3 billion. Excluding military space spending, only a few agencies; such as NASA and the China National Space Administration could realistically support missions of this scale.

If missions with such high costs were introduced today, public backlash could easily create political challenges. This is one of the reasons why humans have not returned to the Moon since the Apollo era, and why the Apollo series of missions was eventually stopped.

This chart represents the approximate costs of the major components of a rocket used historically for major missions like Apollo.

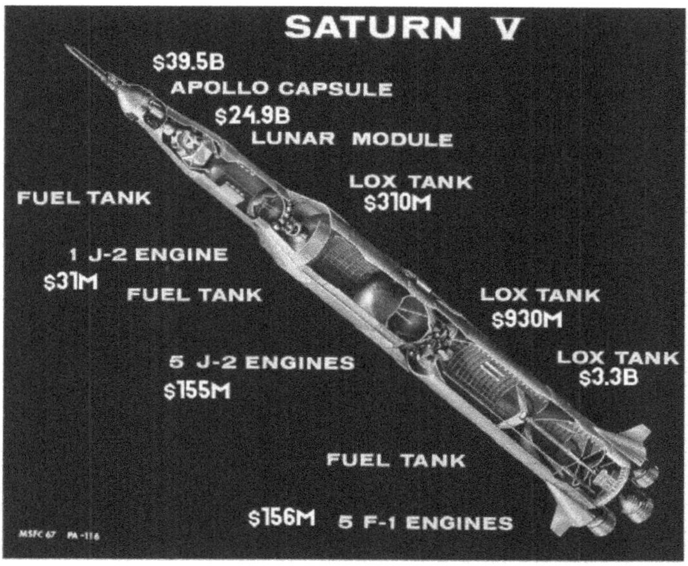

The International Space Station also cost about $150 billion. Although it was built through the cooperation of multiple entities, it was still a very major expense at the time. This is a reality that many would not like to acknowledge.

Today, Indian Space Research Organisation and the China National Space Administration are attempting to build the next generation of space stations. While they may be able to develop the required technology, the cost of assembling a space station and successfully docking each module in orbit is extremely high and involves significant risk.

Such projects can become financially unviable without large-scale cooperation. However, the current geopolitical landscape does not always allow the level of international collaboration that projects like these may require.

Part 4: The Future

Now that this book has covered the past and the present, we would be right to make assumptions about the future of money flow in space. This will be the part of the book where I do not criticise, but instead give solutions to the obstacles. By the future, I refer to the year 2050. In this section, I would like to cover four missions of immense importance to space development.

1. **Humans returning to the Moon**
2. **Human settlement on Mars**
3. **Space stations**
4. **Natural resource extraction in space**

I. Humans Returning to the Moon

For a long time, humans have diverted their focus from the Moon towards other things. However, the spotlight is again pivoting towards one of our nearest celestial bodies. The **Mengzhou, Oryol, Starship, Orion, Gaganyaan, Lanyue, Blue Moon, Starship HLS, HAKUTO-R, Chang'e, dearMoon, ILRS,** and **Artemis** are a few missions intended for the next step of humans on the Moon. The most prominent of them is the Artemis Program, targeted to bring humans near the Moon in the next few days after a series of constant delays. The Artemis Program is dedicated to a special $100 billion, making a possibility viable that the United States will again be the only country to land humans on the Moon after nearly 70 years.

This is a problem to basic economics. A mission which used to cost a total of $10B per flight is now needing a re-initiation cost of 1/10 of a trillion. We need to stop thinking of space as a one-time cost. Instead, think of it as a chain of development, which once initiated will not cost much. However, the larger masses protest over small costs which later cost this much and lead to a larger loss, both in time and progress.

AI is challenging traditional jobs. Space becomes a sector where AI cannot take jobs, but instead act as a supporting hand.

A human mission on the Moon would start a chain of reactions, like a domino effect. The step on the Moon could in turn light a fire for Mars, which is being extinguished by some. It would help gain public support for other well-awaited missions.

II. Space Mission to Mars(Human Settlement)

This is the mission that has been given the highest media attention for the past few years. It all began with a statement of Elon Musk's ambitious dream, and the idea spread like fire. If human landing on the Moon is successfully completed, this would be the next project. NASA, as mentioned in the previous section, is actively spending on the same technology and is the leader in the sector. SpaceX has a massive launch success rate and, along with Musk's dream, is also playing a major role, and the mission to Mars could be a joint collaboration between these two.

Although China is advancing fast, it is unlikely for them to be the first, especially till the magical year, i.e., 2050. Building a human settlement in space would not be an achievement for a country or a company; it would belong to the whole lineage of Homo sapiens. It would impact our day-to-day lives as well, in a positive manner.

Environmental dangers, like the massive data centres necessary for handling heavy AI traffic and processing (and its cooling), could be shifted to Mars. Many resources could also be extracted, leading to a cleaner environment on Earth, if we manage to push mining companies out of Earth. The precious liquid on Earth, water, would be saved, and another safe spot for humanity could be created.

Space tourism could become a reality and human life would become five times better, at the minimum. Like the telephone transformed the world, the same effect would be noticed. Therefore, we need to support it, and not condemn it. Sure, raising the right questions is never wrong, but challenging development is.

III. Space Stations

The math of space stations is complicated. Over two decades from today, the International Space Station was built. An astonishing $150 billion USD was split among a range of countries. However, the landscape has drastically changed now. Countries are at war, geopolitical tensions are high, the world is nearing a massive war, and cooperation in space is slowly dying.

Countries are building stations separately, which only some are capable of supporting. China and India are trying to be the next to establish their stations in space. Russia is too involved in geopolitics and military infrastructure, and the United States is confident in its existing base. The Tiangong Space Station cost was mentioned previously, and the Bharatiya Antariksh Station is estimated at about $2.4 billion. Participation of multiple countries in space is an essential factor in space development, but division in cooperation is a concerning factor.

Space stations play a crucial role in pushing space research on a daily basis and sending humans to space more frequently on a ready-to-use, safe, and trusted platform. The next major space station would likely be from China, because of their massive fund allocation and readiness. India is yet to proceed with Gaganyaan, and without taking astronauts to space independently, development of a space station is not possible so early.

But I do not believe India lacks the technology or money; it is more about capability and other necessities. In a developing economy, allocating funds to such projects is a risky political move that could lead to backlash. Therefore, the next space station shall likely be made by China, followed by India, with a very low possibility of any other country building one before 2050.

> ***"Mystery creates wonder and wonder is the basis of man's desire to understand."***
>
> ***- Neil Armstrong***

IV. Natural Resource Extraction in space

Space is enriched with RREs (Rare Earth Elements) waiting to be extracted, as mentioned in the previous chapter. But then, why haven't we done it yet? The reason is a single major problem-not the material, not the technology, and not the potential, but the cost of transportation itself.

In theory, we could extract resources from an asteroid and return them to Earth within about a decade. However, the process itself is extremely expensive to conduct. Simply transporting the material back to Earth could make it economically worthless because of the high transportation costs.

However, we have not given up yet. Many solutions have been proposed, and several are currently being developed. SpaceX has developed reusable boosters, reducing launch costs from around $10,000 per kilogram to nearly $100 per kilogram. Upcoming systems such as Starship aim to reduce this even further, potentially reaching $10–100 per kilogram.

Another concept gaining attention is swarm robotics. The idea is to send hundreds of smaller robots instead of one large machine. If one fails, the others can continue working, allowing the mission to proceed efficiently while reducing overall risk.

Space sharing is another option. When launching extraction machines, private payloads could be carried alongside them, allowing multiple missions to share a single launch and thereby reducing costs. This could make such projects more economically viable.

The mission is entirely possible if we advance quickly enough and continue progress at the same or even a higher pace. However, there is little room for reversal once such systems are established. We must approach this carefully while also taking advantage of the opportunities it provides.

We should not exploit these resources recklessly, but instead use them as a tool to protect our planet Earth and sustain our way of life without causing harm to it.

The economics of space is slowly changing, a shift that is rarely covered by mainstream media. Therefore, I have summarized the current situation of how money flows in the space sector. I hope you understood it, as it will help you better understand the mission descriptions in the later chapters.

CHAPTER THREE

THE PHYSICS THAT MADE SPACE TRAVEL POSSIBLE

Have you ever wondered where physics formulas help us in real life? We all know that they explain the phenomena of nature, but are they just that, or part of a larger, complex, and advanced system?

The science of launching rockets is not based on a single formula. Instead, it involves a series of complex formulas applied in calculations, determining every drop of fuel and every living being sent into outer space.

However, I have summarized the science of launching rockets in this chapter in a simple and easy-to-understand way. This will help you understand the aeroscience behind rocket launches better than most people.

THE SCIENCE BEHIND ROCKET LAUNCHES

In a highly summarized way, space travel works by launching powerful rockets to overcome Earth's gravity and then using the laws of physics to navigate through space. These processes primarily rely on Newton's laws of motion; the same principles that might have once sounded useless to you.

STEP 1: The Departure

1.The Ignition:

The liftoff of a rocket and even planes is based on Newton's third law, which states every action has an equal and opposite reaction. When Gases heated at high temperatures are thrown by the rocket, the gases apply a equal force on the rocket, driving it forward. It is one of the reasons why a rocket can fly in vaccum without any air, because it pushes the gases out and they push the rocket in the opposite direction. The Thrust force generated by the push of gases is higher than earth's gravitational pull, forcing the rocket upwards. Its basically like to liftoff the rocket, throw gases in the opposite direction.

Action: Throwing of Gases **Reaction:** Equal force is applied on rocket in the opposite direction.

Formula: F=ma

2. The Momentum Chain

The faster the gases are thrown out, the faster the rocket moves. The momentum gained by the rocket must perfectly match the momentum lost by the gases.

$$p=mv$$

A rocket burns propellant (fuel + oxidizer) in a combustion chamber, creating high-pressure gases. These gases are ejected backward at very high velocities through nozzles, imparting a large amount of momentum.

3. The Efficiency Limit: Specific Impulse (Isp)

This is like the mileage of a rocket. It is a number that indicates how many seconds a rocket can produce one pound of thrust using one pound of fuel.

A higher specific impulse (Isp) indicates that a rocket can travel a greater distance using less fuel. It is a measure of engine efficiency and is one of the most vital aspects of aerospace engineering.

4. The Fuel Tax: The Rocket Equation

This is a complex mathematical and efficiency problem in space, because of which we cannot simply go anywhere we want. Each gallon of extra fuel requires even more fuel to carry it.

It is similar to carrying lunch. You need energy to carry it, but to get that energy, you carry snacks; which also require energy to carry. It becomes an endless cycle of energy transfer (as energy cannot be created or destroyed).

This is why rockets mostly consist of fuel, with only a very small percentage made up of other materials such as astronauts or satellites. Every additional unit of fuel requires even more fuel to carry it. At a certain point, carrying extra fuel is no longer worth it, as it demands even more fuel to support itself.

This problem is commonly known as the Tsiolkovsky rocket equation, often referred to as the "tyranny of the rocket equation." You can explore this topic further, as it is a very detailed and interesting problem.

5. The Atmospheric Wall: Max Q

- Max Q (Maximum Dynamic Pressure) is the point during a rocket launch at which the vehicle experiences the maximum aerodynamic structural load or stress. It is a major milestone for a rocket to pass through such high-pressure conditions and is therefore often referred to as the "atmospheric wall." This happens because the combination of decreasing air density and increasing vehicle speed creates the highest drag and external pressure on the rocket's structure.

Why it matters:

- If a rocket were to maintain full speed during the Max Q phase, the resistance and drag could cause it to break apart into multiple pieces. The rocket's structure must be engineered to withstand this maximum force, which typically occurs at an altitude of approximately 11 to 14 km (36,000–46,000 feet), about 60–90 seconds after liftoff.
- To handle this, engineers reduce the engine thrust to around 60–80%, lowering the rocket's velocity and helping it cross Max Q smoothly

without damage. After passing this point, the engines can return to full power and continue the journey forward.

STEP 2: The Capture(Transition to Orbit)

1. The Sideways Fall: Centripetal Force vs. Gravity

An orbit is the result of a series of forces acting together. It is like being pulled downward while moving in a circle. In simple terms, an orbit is a perpetual free-fall where an object moves sideways so fast that it "misses" the ground.

Imagine you are standing on a tall mountain. You launch a cannonball at a low speed; the ball travels a short distance and lands on the ground. If you fire the cannonball at a medium speed, it lands farther away because the Earth's surface begins to curve away beneath it. However, if you launch the cannonball at orbital speed, the Earth's surface curves away at the same rate. The ball is still "falling," but it never gets any closer to the ground.

An orbit is a perfect balance between inward force (centripetal force), gravity (the downward pull), and inertia (the tendency of the object to move in a straight line). It is a delicate equilibrium achieved through complex physics and careful study of the motion of the Earth, the Sun, and other celestial bodies.

2.The Gravity Grip: Escape Velocity

Escape velocity is the speed required to break free from the gravitational pull of a celestial body. An important distinction to note is that it is not the speed required to simply reach space, but the speed needed to never return to the body again without using any additional fuel.

For example, the International Space Station is in space, but it is not at escape velocity. The Earth acts like a deep gravitational well, and for an object to escape this well, it must have greater kinetic energy than its gravitational potential energy.

From Earth, this speed is approximately 11.2 km/s, which is extremely high, especially considering it is in S.I. units. This converts to about 40,320 km/h or 25,000 mph. If you observe closely, you may notice something interesting: the circumference of the Earth is also approximately 40,000 km. There seems to be a connection between these numbers.

So, what is it?

Well, escape velocity is not dependent on the mass of the object trying to escape, but on the mass and radius of the celestial body itself (which determine its gravitational pull). Therefore, a rock and a spaceship would theoretically require the same escape velocity to leave the gravitational field of the same celestial body.

Interplanetary and deep-space missions must overcome Earth's gravitational hold, making it extremely important to calculate these values accurately. Space missions are expensive and carry a nation's pride and ambition. Even a small mistake can lead to major setbacks, which is why precision in space exploration is essential.

This topic has been discussed in greater detail in my previous book, From Earth to Mars. Please feel free to explore it if you have time.

(Though the correlation between Earth's escape velocity mentioned above appears to be a coincidence, it can be observed in certain cases. However, the difference increases significantly with changes in the size of planets, using Earth as a reference.)

3. The Timing: Kepler's Third Law

This law, in very simple language, states that smaller orbits take less time, while larger orbits take more time. It works together with many other principles to make space travel possible.

$$T\text{\^{}}2 \propto a\text{\^{}}3$$

Engineers use this to calculate the correct altitude for satellites and space stations, the motion of asteroids, and many other complex factors. A real-life application can be seen in communication and weather satellites.

Suppose you want a satellite to remain over the same location on Earth at the same time every day. This law shows that at an altitude of about 36,000 km, the orbital period is approximately 24 hours; the same as Earth's rotation. Hence, the satellite can consistently cover the same region.

For missions that involve traveling between planetary orbits, this law also helps engineers plan efficient and shorter routes.

STEP 3: The Way(Moving through Void)

1. The Speed Boost: The Oberth Effect

As mentioned in the previous edition (From Earth to Mars), an orbit has a highest and lowest point. The lowest point is referred to as the periapsis.

The Oberth effect states that firing a rocket engine when it is already moving at high speed makes it significantly more efficient. Burning your engines at higher speeds provides much more useful energy compared to burning them at lower speeds.

This is very useful in maximizing efficiency in space and saving fuel for longer journeys. When a rocket is transitioning out of its current orbit toward another celestial body, it reaches its highest speed at the periapsis (the lowest point). At this point, the engines are fired to gain maximum benefit and save energy during transit.

Imagine throwing a ball while standing still versus throwing it while sprinting. The ball will travel faster in the sprinting case because it carries the kinetic energy of both the sprint and the throw.

2. The Fade Out: Inverse Square Law

It is a crucial law for establishing communication and safety in space missions. It states that when an object moves away from a source, the intensity or strength of a physical quantity changes; not directly, but by spreading out over a larger area. As the distance increases, the intensity spreads across a greater region of space, causing it to decrease.

$I \propto 1/d$ (square)

This principle applies to gravity, sound waves, and communication signals, all of which are important for safe navigation, especially in missions carrying human lives.

Imagine a sphere around a rocket. As the distance increases, the sphere grows larger. The same amount of energy is now spread over a much bigger area, which is why the intensity decreases.

It is also useful in applications such as solar panels, where measuring how much solar energy can be intercepted is important for efficiency and innovation.

STEP 4: The Survival (Physics inside the rocket)

1. The Heat Trap: Stefan-Boltzmann Law

The Stefan–Boltzmann Law lays down the principles for developing livable conditions in a rocket and making the temperature bearable. It is used to calculate the heat on objects during a specific period and then build technologies accordingly to handle the immense strength.

In space, there is no air to cool down the ship. The only way to get rid of heat is to radiate it as light (because energy cannot be destroyed, only changed)(Light hereby refers to several forms of light which includes electromagnetic energy also). The radiators must be big enough, or the crew could potentially die.

The area in which astronauts live is compact, and even a small rise in temperature could take a dangerous turn. The space suits are also modified accordingly, with the outer layers acting as reflectors, and even a little damage could cause instant consequences, something which is not manageable, especially with limited resources and little room for movement.

2. The Pressure Stress: Hoop Stress

There is air on Earth. The air puts pressure on the Earth. Now, the air inside your body and outside has the same pressure, and they balance out. Space, which is a vacuum, has no air pressure.

However, the human body is habituated to a certain air pressure, which needs to be maintained for survival, like it is on airplanes. Therefore, the capsule needs to be pressurized to keep the astronauts alive in livable conditions.

But the pressure outside the capsule and inside is largely different. Because of this, the air inside the capsule pushes on the walls, and the walls could get damaged and lead to a tragic incident. The metal wall of the capsule acts like hoops, holding it together. Thicker walls act as the only protection in such a situation.

Note that this also acts as an extension of the mathematical problem mentioned previously, where we cannot use very heavy metal because it would require too much fuel. A wider ship with the same air pressure as a smaller ship would face more stress, which is why building large stations like the International Space Station is a big challenge, and many shuttles are built compact.

This physics is very important for any object traveling above 10,000 feet carrying human lives, because the air above that is not very stable and could lead to serious health risks, and even death within minutes if necessary action is not taken.

STEP 5: The Return(Energy Change)

1. The transition(Kinetic to Thermal)

In order to land, the rocket must reduce its orbital speed completely. But energy cannot be destroyed. Therefore, the rocket hits the atmosphere of the celestial body at high speeds, which causes friction. This, in turn, helps convert the energy into thermal energy (heat).

$$E\_k = (1/2)\ m\ v\text{\^{}}2$$

The atmosphere acts as a giant brake, converting the rocket's motion (kinetic energy) into heat (thermal energy) to slow it down for landing.

2. The squeeze: The adiabatic Compression

After the rocket experiences friction, it already heats up. But this isn't even the beginning. The rocket hits the air so fast that the air doesn't have time to move and gets compressed, and this compression, according to the laws of gases, heats the rocket further.

Specialized heat shields (ablative or ceramic tiles) are used to absorb or dissipate this heat, preventing the rocket's structure from melting. The air compresses and heats rapidly because the capsule travels at speeds of 17,000 mph+ (orbital speeds).

3. The Final Balance: Terminal Velocity

A perfect balance is needed to land an object safely. When an object achieves a uniform speed, it is called terminal velocity. A capsule falling from space initially speeds up due to gravitational pull. However, after some time, the drag increases to such an extent that the velocity stops increasing, and the object does not accelerate further.

$$mg = \frac{1}{2} \rho C\_d A v^2$$

Each and every object has its own terminal velocity. Heavier objects have higher terminal velocity, while a larger surface area increases drag, lowering the velocity.

It is important for engineers to calculate and ensure that rockets reach lower terminal velocities to achieve a smooth and controlled landing rather than a strong, unstable one. Once terminal velocity is reached, parachutes are opened to reduce it further, and then the capsule is safely landed, preferably on less compact surfaces like water to soften the landing and avoid damage before rescue.

This was the summary of the physics involved in a spaceflight. However, this isn't the whole science behind it because a rocket has lakhs of calculations involved, and it is too boring or not relevant to study all of it. If you are interested, I have sourced some of it in the bibliography. You can research more on it. Thanks!

The Perseverance Rover

Perseverance is one of the larger and more recent rovers deployed on the surface of Mars. It is big in size, similar to a normal car. It was launched in the year 2020 by NASA and was similar to the rover Curiosity, but more technologically advanced and useful.

This rover was largely built and designed at the Jet Propulsion Laboratory, Pasadena, in the American state of California. After it was built, carefully keeping in mind the harshest of conditions, it was safely packed and moved to the Kennedy Space Center in Cape Canaveral, Florida. This is the location from which many of the most important launches in the history of mankind have been conducted.

The components were loaded into the payload section at the top inside an Atlas V rocket. The rocket was operated by the United Launch Alliance (ULA). The Earth and Mars align for a suitable launch every 26 months. As we know, Earth takes 1 year to complete an orbit around the Sun, while Mars takes almost 2 years.

A launch is conducted when Mars and Earth are nearly aligned, and this is known as a launch window. Perseverance was launched on July 30, 2020.

The four solid rocket boosters used all their fuel first. At 1 minute and 49 seconds, they were separated from the body of the rocket at an altitude of 145 km. At 3 minutes and 27 seconds, the payload fairing was separated at an altitude of 391 km. All these separations of now unnecessary components increase the efficiency of the rocket. Soon after, the main booster ran out of fuel and was separated at 4 minutes and 28 seconds at an altitude of 756 km.

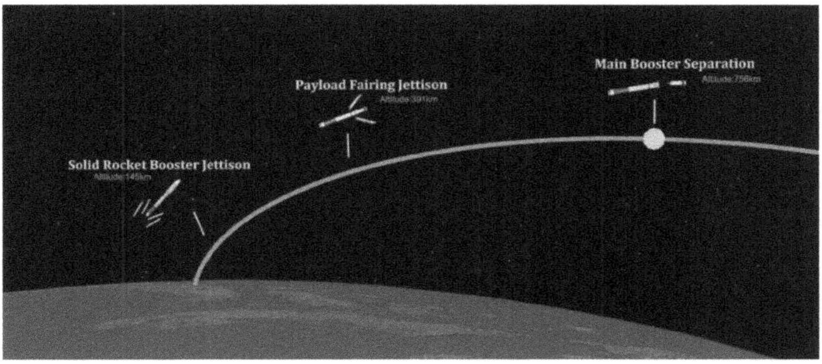

The next stage, the Centaur stage, is now activated and its main engines fire up. This is the part which helps us achieve an orbit around the Earth. At 11 minutes and 30 seconds, the engines are turned off. The coast stage initiates for about half an hour, where the rocket floats in orbit, and it is ensured that everything is still in check.

After that, the engines are re-ignited for about 8 minutes to escape the Earth's orbit (known as the Earth escape burn). This helps us get a trajectory towards Mars. The spacecraft then separates at an astonishing speed of 41,000 km/hr, and it takes about seven months to reach Mars.

The spacecraft has a few parts. The payload is placed between a backshell and a heat shield for safety, and behind the backshell is a cruise stage, which has power, thrusters, fuel tanks, and maintains communication with Earth.

The path adjustments along the way are carried out by small but powerful thrusters, which consume little fuel but are very efficient. This is called a Trajectory Correction Maneuver (TCM). Several of these are carried out along the way, as we cannot achieve a perfect trajectory initially.

Landing on Mars is hard. It is known by the acronym EDL (Entry, Descent, and Landing), nicknamed the "seven minutes of terror." If a single error occurs during this period, it is almost certain that a crash will occur.

This is because, at the time of landing, it takes about 11 minutes for signals to reach Earth, even at the speed of light. This is a very long time. For perspective, from Earth to the Moon, communication takes just over a second to reach. When Apollo Moon landings were happening, NASA could communicate in near real time. It would take about 2.5 seconds to receive a reply.

It is a completely different situation on Mars. If a problem occurs during landing, mission control would only hear about it after 11 minutes, and sending new commands would take another 11 minutes, leaving almost no chance of saving the mission. This means everything must be self-sufficient and ready to face any challenge, and nearly perfect to carry out such a high-risk mission.

Now let's cover EDL step by step.

The cruise stage detaches about ten minutes before entering the atmosphere, at about 1600 km from Mars, because it is no longer needed. Small thrusters on the backshell are then used to maneuver the spacecraft.

Next begins atmospheric entry at an altitude of 122 km above Mars. The heat shield protects the spacecraft during this phase.

At about 11 km above the surface, parachutes are deployed using a technology called range trigger, which detects how far the spacecraft is from the ground and deploys the parachute at the right time.

Next is heat shield separation at about 9.6 km above the ground, and the Perseverance rover can finally see the surface.

The rover then takes images of the ground and matches them with its onboard data in a process called terrain relative navigation. Using this, the rover is guided safely toward the intended landing site.

At about 2.2 km above the ground, the backshell is separated. All that remains is the payload, which contains the descent stage and rover.

The descent stage fires its engines to slow the rover as it approaches the ground. Then the rover separates but remains attached by cables, like a crane, at about 20 meters above the surface. This is called the sky crane maneuver.

Finally, after this entire sequence of events, the rover touches the Martian surface and detaches from the descent stage.

This rover was very advanced and helped us prove that the calculations and laws we implemented actually work. This will help us build rockets that work almost definitely, with very little chance of error.

Components of Perseverance

Physics contributes to a lot in our lives. However, it plays an even bigger role in advancing what is yet to come. What we couldn't have imagined has become a reality in the past century. From airplanes to rockets to humans on other celestial bodies, it was all once considered impossible or just a dream, but it is a reality today solely because of our commitment to progress and physics.

We cannot possibly test everything in space, as it costs too much, but we can use physics to predict outcomes accurately and work on them. **Before we reach space, physics reaches there first.**

PART II

HUMANITY'S JOURNEY SO FAR

CHAPTER FOUR

BEYOND THE BLUE PLANET

What have we really achieved in space? This chapter is not negative, but rather positive; and one of the most important of all. It will cover the missions we have conducted till date, what we have achieved, and what we still have to.

Out of the infinite void, have we done something even significant; something that someone can look upon and call us a decently advanced civilization? This is the truth, and it is not as harsh as you might believe or what the media portrays.

You might have heard of only a few missions; Apollo, Chandrayaan, Mangalyaan, Gaganyaan, NASA's rover series, Sputnik, and some others. However, the reality is much larger and better, we have achieved a lot.

We need to celebrate and understand what we have accomplished, and what is still left to do.

Missions in the Past

The Major Ones

-Apollo Program

The Apollo program is probably the most important and well-known series of missions in the world. It broke not 1, 10, 50, or 90; but hundreds of records. It achieved what we couldn't for thousands of years. It was like a dream come true.

- This is just a list of a few records they broke:
- Fastest manned flight
- Lowest altitude over the Moon
- First crewed flight of the lunar module in lunar orbit
- Longest distance from Earth (at the time)
- Longest time in lunar orbit
- And hundreds more

I would like to cover Apollo 17 in this book, a lesser talked of missions with ofcourse the same specifications as all of them had.

Apollo 17

- Each mission was launched from Cape Canaveral, Florida, and mission control was in Houston, Texas. There stood a Saturn V rocket, which was 363 feet (110 meters) tall.

The rocket was divided into three stages, with each stage referred to by a technical name. The first stage was called the S-IC, with five F-1 engines. The second stage was the S-II, powered by slightly smaller J-2 engines. The third stage was the S-IVB, powered by a single J-2 engine.

- The actual spacecraft was stored in the upper stage, which comprised the Lunar Module (LM), which would land on the Moon. The Service Module had the power to enter and leave lunar orbit and supported other systems. The Command Module was where the astronauts spent most of their time. At the very top, there was a Launch Escape System (LES), which would help the astronauts escape in case of an emergency during launch.

The rocket detached from the tower beside it and launched within seconds. As the rocket gained speed, the astronauts faced up to four times Earth's gravitational pull, though they were well trained for such conditions.

- The S-IC engines cut off at about 2 minutes and 42 seconds after launch, at an altitude of 67 km. The second stage then fired, and the first stage and LES separated. The second stage shut down at 9 minutes and 12 seconds at an altitude of 175 km.

The third stage fired until 11 minutes and 39 seconds at an altitude of 190 km to achieve parking orbit, but was not yet removed. After orbiting Earth 2–3 times and ensuring everything was functioning properly, the third stage was re-ignited for about 6 minutes for Trans-Lunar Injection (TLI), placing the spacecraft on a path to the Moon.

- The S-IVB stage was then discarded. What remained was the most important section of the spacecraft. The Command and Service Module (CSM) rotated 180 degrees to dock with the Lunar Module and extract it from the S-IVB.

At this point, the flight path resembled a large elliptical orbit around Earth (if the Moon is ignored). As the spacecraft approached the Moon, a free-return trajectory was achieved. This meant that if something went wrong, the spacecraft could return to Earth without major corrections.

- After this, Lunar Orbit Insertion (LOI) began. Engines were fired in the opposite direction of motion to slow the spacecraft and allow it to enter lunar orbit. The Service Module engine fired for about 6 minutes for this maneuver.

When ready, two astronauts moved into the Lunar Module while one remained in orbit. The LM separated from the CSM and headed toward the Moon. It first moved about 2 miles away for safety, then fired its engines for about 30 seconds for Descent Orbit Insertion (DOI), bringing it to about 50,000 feet above the lunar surface.

- The engine fired again for Powered Descent Initiation (PDI). After a series of maneuvers, the LM achieved touchdown.

The astronauts then descended, conducted scientific activities, and prepared for the return journey. The Lunar Module was divided into two parts: the descent stage, which was left behind, and the ascent stage, which would carry the astronauts back.

- The ascent stage fired for about 7 minutes to re-enter lunar orbit. Several short burns were conducted over the next 3.5 hours to reach a docking point with the CSM. After docking, valuable materials were transferred, and the LM was detached permanently.

The Service Module then fired for about 2.5 minutes for Trans-Earth Injection (TEI), beginning the 3-day journey back to Earth.

- Once near Earth, the Service Module was discarded, leaving only the Command Module. Re-entry was the final and most intense phase. Earth's atmosphere generated extreme heat, but the heat shield protected the capsule. Parachutes were then deployed, and the astronauts were safely recovered by ships or aircraft carriers.

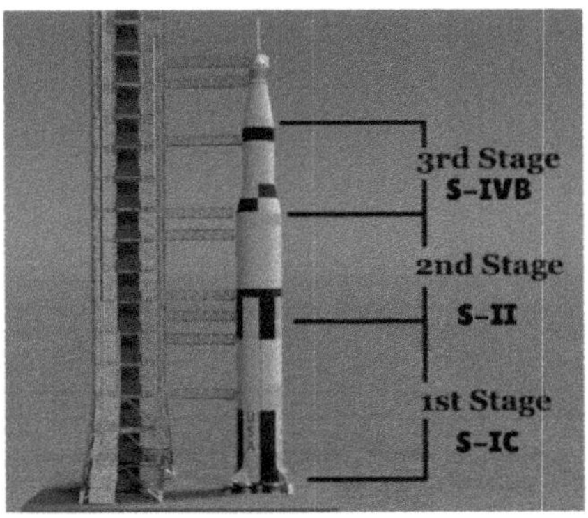

i-

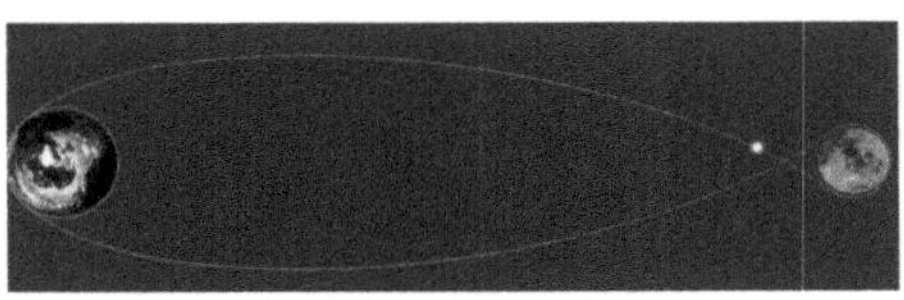

Free Return Trajectory

-Hubble Space Telescope

The **Hubble Space Telescope** was one of the most advanced telescopes we have ever produced. Launched on April 24, 1990, it helped us capture images that had the power to explain phenomena we had never imagined exploring. From pictures that could help explain the beginning of the universe to a mistake that nearly made it a failure, it faced it all.

It consumed relatively little electricity, comparable to a hairdryer, while using the largest source of energy in our solar system; the Sun. The Hubble Space Telescope has a mirror that is about 8 feet across. It is around 44 feet long and weighs more than 25,000 lbs.

It travels at about 10,000 miles per hour around the Earth, completing one orbit in roughly 95 minutes. It is fast, precise, and phenomenal.

However, Hubble is different from other telescopes. It was the first of its kind that could be fixed, repaired, and upgraded by astronauts. Astronauts could visit it, repair it, and install better instruments without the need to launch an entirely new telescope. This significantly extended its life and capabilities.

It has made over 1.8 million observations and is still more powerful than many ground-based observatories. However, it is now on the brink of being decommissioned.

Discoveries by Hubble

-In the year 2011, the Hubble Space Telescope achieved two milestones. It made its 1 millionth observation and contributed to its 10,000th research paper.

- **Discovery No.1** - In the year 1994, scientists used the Hubble Space Telescope to explore the Messier 87 galaxy, which is much larger than ours. Through this research, we understood that most galaxies have a giant black hole at their center, playing a major role in their structure.

  However, there is also something called dark matter. It is not clearly visible, but it exerts gravitational influence throughout galaxies, playing a significant role in holding them together.

Discovery No.2 - Any telescope before the Hubble Space Telescope, whether in space or on land, could detect exoplanets but could not study their atmospheres in detail. Hubble, unlike others, had advanced technology to identify certain characteristics of a planet's atmosphere.

A planet being habitable depends largely on its atmosphere, because a suitable atmosphere can support life. In the year 2001, Hubble studied the HD 209458 b and detected elements in its atmosphere using light, without physically traveling to the planet.

It does this by analyzing how light passes through the planet's atmosphere. This technology has helped scientists study atmospheric composition and even understand weather patterns on distant planets.

- **Discovery No.3** - In the year 2004, the Hubble Space Telescope released a truly fascinating image. The image showed about 10,000 galaxies in a single frame. This image is known as the Hubble Ultra Deep Field.

  This led to a wave of discovering new galaxies.

  If you stand far from a train, you will likely hear less of the sound of the train's horn, right? In space, there is a similar phenomenon where, as light travels farther, it shifts toward the red end of the spectrum. This is called the Doppler effect (more specifically, redshift).

  Scientists used this to measure the distance of faraway galaxies. This all led to a striking discovery; or rather, strong evidence for an existing theory; that the universe is expanding. It was observed that galaxies are moving away from us, which means that the universe is growing.

Discovery No. 4 - This is a very major discovery in the field of science. It basically shows that gravity can act like a magnifying glass. How does NASA get images of distant galaxies so clearly?

Imagine a galaxy behind a foreground galaxy. The distant galaxy emits light in all directions, and the foreground galaxy's gravity bends that light, similar to a convex lens. This effect is known as gravitational lensing.

This bending of light creates a magnified and sometimes distorted image, which is then captured by the Hubble Space Telescope.

This list doesn't end here, but these are some of the most groundbreaking and well-known discoveries. You may research them further if you are interested. The Hubble Space Telescope was truly a wonderful and remarkable telescope built by mankind.

-The Curiousity Rover

- The Curiosity rover was one of its kind and was sent after the Opportunity rover started showing signs of weakness. It was launched on 26 November 2011 and landed on Mars on 5 August 2012.

The rover landed in the Gale Crater and is still operating there, finding evidence related to past life. Its main purpose was to study the history of Mars and find if life had been possible there in the past.

- The plan was to find similarities between water deposits on Earth and Mars, and whether water could have existed there. When water flowed into the crater, the deposits must have settled down, and those were studied. The rover found strong evidence that the crater was once filled with water.

In order to investigate further, Curiosity dug into the lakebed and found organic matter. Organic matter are chemical compounds containing carbon, which is found in life on Earth. This is a sign that microbial life could have existed on Mars thousands of years ago.

- The rover has travelled about 22 kilometers on the Martian surface and continues to move further. The tires on Curiosity have suffered a lot of wear over its lifetime, and scientists divert its path from time to time to reduce this damage.

Curiosity remains in contact with Earth using the Deep Space Network. Large antennas are located at three places on Earth: Madrid (Spain), Canberra (Australia), and California (USA). It can send signals directly to Earth using its high-gain antenna, but this process takes a lot of time and battery power.

- Therefore, the rover often uses the Mars Reconnaissance Orbiter, which passes over it every day for about 8 minutes. Around 30 megabytes of data can be transferred during this time.

The rover uses a Radioisotope Thermoelectric Generator (RTG), which converts heat from radioactive isotopes into electricity. This power is enough to charge its batteries and allows Curiosity to carry out high-power tasks daily.

- The RTG is expected to power Curiosity at least until 2026, i.e., this year. After this, its functions may become limited as the power supply slowly decreases. However, power might not be what ends Curiosity; Mars itself could. The dust storms that ended Opportunity could also affect Curiosity.

Scientists may slowly phase out Curiosity by turning off its instruments one by one to preserve power, until it sends its final signal. The rover is currently climbing Mount Sharp, which is about 5 km tall. Its tires are already heavily damaged, and we cannot expect it to last for much longer.

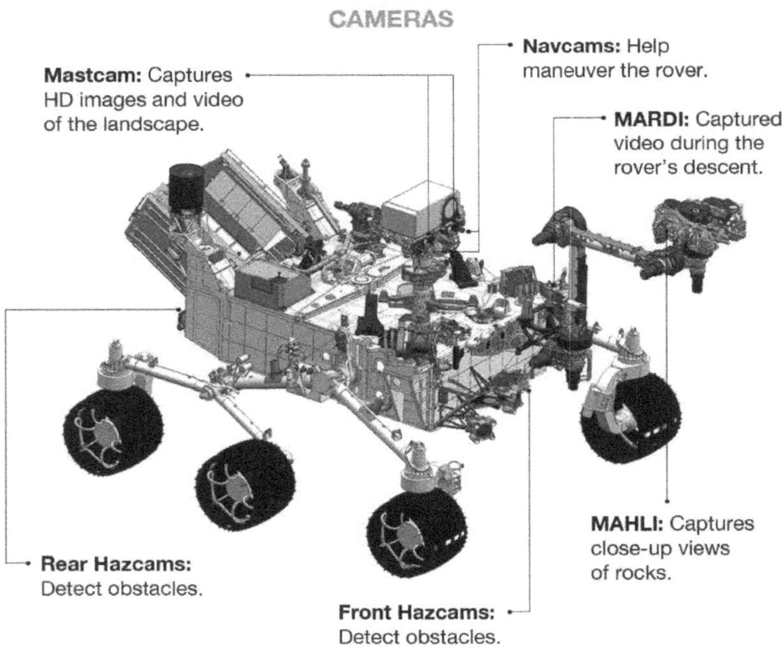

-Cassini–Huygens

The **Cassini–Huygens** mission originated about four decades ago. It was a joint European–American mission to investigate Saturn, its rings, and everything around it. It was launched on 15 October 1997 and took about 7 years to reach its desired orbit.

- Saturn is about 10 times farther from the Sun than Earth is, which is roughly 1.5 billion kilometers from Earth.

Gravity assist was an essential part of this mission and many others. It is the process of using the momentum and gravity of other planets to push the spacecraft forward without using much fuel. Faster and longer missions use it to become more viable and achieve higher speeds.

- The spacecraft looped around the inner solar system twice, which took about 2 years. It flew by Venus, then Earth, which put it on a path toward Jupiter by the end of 2000. That set it on the final path toward Saturn. It achieved orbit around Saturn in the year 2004.

Cassini gave us images we could not have achieved with simple flybys. We learned about the processes that occur in the rings of Saturn. These processes mimic how planets formed in the protosolar nebula and mirror the dynamics of massive spiral galaxies.

- On December 5, 2010, a major storm was observed as it developed and evolved until its end. This helped scientists better understand meteorology on large planets and compare those patterns with Earth.

It told a story of how systems evolve; the processes that created Saturn's moons, with its rings acting like a miniature solar system. The processes through which Saturn's moons were formed are likely similar to those that led to the creation of our own solar system and planets.

- There were a few developments made on one of the moons of Saturn, Enceladus. This moon was discovered to have a global subsurface ocean of liquid water, along with plumes of water ice, salts, and organic molecules erupting from fissures at its south pole.

This could mean that conditions on this moon may be suitable to sustain life. Evidence was also found for hydrothermal vents on the seafloor of this moon, suggesting potential microbial life.

- The European Space Agency (ESA) Huygens probe was carried by the Cassini–Huygens mission, which later successfully landed on Titan in 2005, providing the first direct images from a moon in the outer solar system.

This mission revealed a lot about Titan's atmosphere and surroundings. It showed that Titan has a methane-based weather system, with rain, rivers, lakes, and seas of liquid hydrocarbons, remarkably similar to Earth in structure (but not in composition).

- Cassini also captured detailed imagery of a persistent, hexagonal-shaped jet stream surrounding Saturn's north pole.

The mission ended on September 15, 2017, with a pre-planned plunge into Saturn's atmosphere. This was done to safely dispose of the spacecraft and prevent it from crashing into Titan or Enceladus, which could contaminate potential life. These moons are considered top candidates for finding life, and contamination could be extremely harmful.

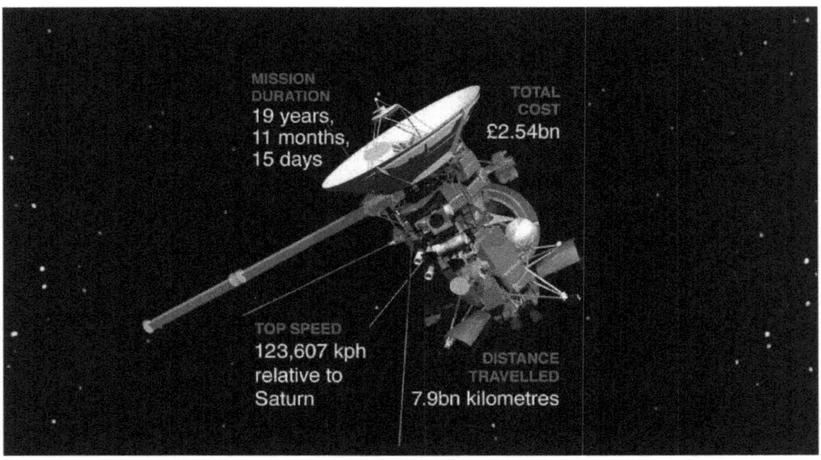

Important

-Sputnik Missions

Sputnik 1 (launched on October 4, 1957) was the beginning of the current era of space. It was the first artificial satellite in space by mankind. It marked the beginning of the space race between the Soviet Union and the United States, and the start of a decade which would provide so much to technology that we had never imagined.

- It was a shiny metal ball with 4 antennas, and was known for its iconic beep. It acted like a radio beacon or transmitter in space, which sent its signals for approximately 21 days. It operated in an elliptical Low Earth Orbit (LEO) and travelled around the Earth every 96 minutes.

The mission shocked the United States and might have been the start of a "war" that was good, not harmful to anyone. A decade after this mission, we reached the Moon, showing a truly remarkable speed of progress.

- The PS-1 was launched late at night on October 4, 1957. The boosters separated at about 116.38 seconds, and the second stage burned out at 314.5 seconds. Approximately 20 seconds later, a spring ejected the conical fairing of the launch vehicle, and Sputnik was gently pushed into space by an oxygen-actuated piston.

Sputnik was now in an oblong orbit with a low point (perigee) of 228 km and a high point (apogee) of 947 km, with an orbital inclination of 65 degrees. People around the globe with radio sets could hear its beep-beep.

- It continued beeping for 326 orbits, with the batteries running out on October 26, 1957. The frequency of transmission never changed, signifying that the temperature inside remained constant and that it was never breached.

It continued its silent revolutions, eventually falling out of orbit on January 4, 1958, and burning up in the dense atmosphere of Earth. It completed a total of 1,440 revolutions before falling back.

- Soon after the success of Sputnik 1, the Soviet Union continued its journey, with Sputnik 2 launching soon after on November 3, 1957, carrying the first animal in space, the dog Laika. In May 1958, Sputnik 3 was launched into orbit in a D-design, while the United States had already launched Explorer 1 in February 1958.

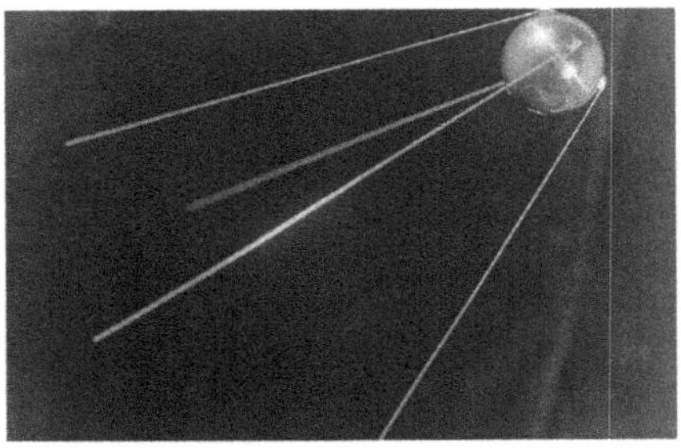

-

-Vostok Program

The R-7 Semyorka was the base of many rockets. The Sputnik program also used the R-7 and successfully completed its journey. It was a modification inspired by the V-2 rocket and was originally meant for military operations, but after the United States announced its ambitious plans, the Soviet Union needed to succeed. The rocket was built with the help of German scientists captured after World War II.

- The Vostok program used a modified R-7 rocket renamed to Vostok-K. The new rocket added a third stage and significantly increased the capacity compared to the R-7. The R-7 had a capacity of a few hundred kilograms, whereas the Vostok-K could carry 4,730 kg of weight.

The mission planned to send the first human into Low Earth Orbit, beyond the confines of this world. Yuri Gagarin was selected for Vostok 1 because of his special skills and recommendations. The cosmonaut boarded the rocket on 12 April 1961, a day that will never be forgotten in history.

- The engines roared to life at 9:07 a.m. (local time), and each and every movement, every signal, and every second was closely watched by the world and the media. Gagarin sat upon 4,725 kg of metal and fuel. The rocket continued its journey in orbit for 1 hour and 48 minutes. The rocket reached a maximum altitude of about 301 km.

The controls of the rocket were automated, as there was a high risk of humans losing consciousness in space and zero gravity, though the controls could still be accessed through a key kept in a sealed envelope.

- There was a problem with the service module separating; the capsule spun before finally separating and re-entering. As planned, Gagarin ejected at 7 km above the ground, landing separately by parachute in a field near the Volga River at 7:53 UTC.

He was awarded the highest honors soon after, but was kept safe and somewhat confined because of potential intelligence risks. This mission marked a huge victory of the Soviets over the Americans in the space race and proved that human spaceflight was survivable and possible with enough effort.

- Mass demonstrations were held all across the Soviet Union, comparable to the victory parades held after World War II. The United States was shocked, and communism had, for the moment, proved itself stronger in this domain.

The aftermath of Sputnik and Vostok led to the creation of NASA.

- The Fédération Aéronautique Internationale (World Air Sports Federation) recognized three records for Gagarin: duration of orbital flight (108 minutes), greatest altitude in Earth orbit (327 km), and highest weight lifted in Earth orbital flight (4,725 kg).

The Soviets kept their plans a secret before their success, only announcing them after completion.

-Skylab

If you were to ask any average person, or even yourself, a simple question,"What is one of the human stations in space?"; you would likely get a simple answer: "International Space Station."

- But years before the ISS, NASA made a space station that is not widely noticed today, though it might be one of the underlying reasons for the success of the ISS. The Skylab missions are among the most negatively perceived missions, but they helped us discover what we couldn't for years.

It had problems from the minute it launched, but it also carried advantages. Humans, before this mission, had spent only a few days in space. However, the next large goal was Mars, and unlike the Moon, which takes about 3 days of travel one way, Mars takes around 7 months. We were yet to assess how human bodies react to zero-gravity environments for more than a year.

- A space station was being envisioned as a hub between interplanetary transfers, and therefore its development was necessary. During the Cold War, the United States prioritized the lunar landing and kept the idea of a space station aside, which the Soviet Union capitalized on in 1971 by launching the first space station, Salyut 1. The United States now had to catch up.

Due to the cancellation of later Apollo missions (18, 19, 20), they had spare Saturn V rockets. The plan was to convert the second-stage fuel tank into a workspace and use crew modules to transport astronauts into it.

- Skylab was launched on May 17, 1973. Within minutes of launch, the heat shield was detached due to a problem. As soon as the station reached orbit, it started heating up because of the absence of the heat shield, and solar radiation further increased the problem.

The next crew launch to the station was delayed because of these issues, and NASA started working on solutions. About 10 days later, the crew was equipped with a solution and was expected to save the $2.5 billion program. This became the first mission in which astronauts would perform major repair work in space.

- They used a covering to shield the exposed area where the heat shield was lost during launch. Hundreds of experiments were carried out during the program, and it finally ended when Skylab burned up and re-entered Earth's atmosphere on July 11, 1979.

This was a huge success despite its problems and helped clear many myths about the effects of zero gravity. The workspace was like a huge aluminum can with a lot of open space, allowing astronauts to perform experiments more freely.

-Venera Program

The USSR was committed to expand further into space and other planets, and to potentially find life and livable environments. Venus was perceived as a planet that was covered with dust and cloud, and beneath lay a world similar to Earth, both in temperature and pressure. To prove this theory, the Soviets announced the Venera Program. The first few missions were, as always, test flights and flybys, testing their capacity and technology.

- Finally, on July 12, 1967, Venera 4 was launched. It had a simple goal: to cross Venus's atmosphere and send back the temperature and pressure readings of Venus. In October of 1967, Venera 4 entered the atmosphere and its parachutes deployed at 40,000 km/hr. In the beginning, the atmospheric pressure and temperature weren't too different from Earth. But as it progressed inside the atmosphere, the readings started going crazy. One atmosphere became 20, and the temperature rose rapidly.

Suddenly, Venera 4 blasted open and lost its connection to Earth. The last reading it gave was: 22 bar of pressure (Earth's is 1 bar), and 280 degree Celsius, 22 times thicker and 250 degrees hotter than Earth. This revealed how harsh Venus really was, and now they were dedicated to exploring Venus.

- The initial design was relatively simple: a 1 meter wide sphere, pressurized to inside 25 bar. It had a handful of instruments and a battery to support it all. The lander didn't have thrusters and used a series of parachutes to drag itself down. However, this was a problem. The lander hit the thick atmosphere like a brick wall, slowing it down rapidly to just 18 km/hr. Because the lander was travelling slowly, it had to withstand the pressure for a longer time and caused significant damage.

The Soviets realized that once the lander entered the atmosphere, time was limited. They found a simple solution: to add a chord around the parachute to stop it from opening completely until it is very close to the ground. The Soviets made the walls of Venera 7 2 times thicker and the pressure to 182 bar.

- In December of 1970, Venera 7 entered Venus. Everything went perfect and according to the plan, until 3 kilometers from the ground, where the parachute tore itself and the approach was destabilized. The probe crashed into Venus at 60 km/hr. However, it was discovered that the lander actually transmitted a few weak signals after, showing the temperature to be a mind-blowing 500 degree Celsius.

To solve this problem, the Soviets kept blocks of lithium nitrate, which had a very high heat diffusion (the temperature needed to convert it into a liquid). This system worked extremely well, and the Venera 8 successfully touched the surface of Venus safely.

- After this massive success, Venera 9 was sent to Venus to take photos of Venus. The design was changed once again. Instead of parachutes, Venera 9 had a large air brake to slow the lander down during the final descent and landing legs for a safe landing. It had 2 movable cameras which could take a 360 degree panorama of the surrounding area. The technology to survive this atmosphere was yet another massive innovation, using a periscope and movable lens to capture the light without actual exposure.

-Mars Orbiter Mission(MOM)/Mangalyaan

The Mangalyaan Mission, popularly known as MOM in the Western media, was an amazing feat achieved by ISRO (Indian Space & Research Organisation), in which they inserted a probe in a highly elliptical orbit around Mars. This was achieved back in the year 2014, with just 66 million USD. This mission challenged NASA's legacy, though their probe (Reconnaissance) did have higher coverage.

- This made ISRO the first one to get a mission to Mars right on the first try, and the first Asian agency to get to Mars. That itself is a large win, but what else has it achieved in the past 12 years in orbit? Well, it was declared dead by ISRO in the year 2022, because it ran out of fuel and battery after a long eclipse, and all communications were lost.

However, for the remaining 8 years, it provided a lot to us, what NASA and ESA even couldn't. The beautiful thing about this mission was that ISRO put a raw camera in there, providing images that you would see if you went to Mars. These images showed ice caps, carbon dioxide atmosphere, and the fascinating surface of Mars.

- In the imagery captured by Mangalyaan, you can clearly see a lot of volcanoes on the surface of Mars. You can also see the dust storms that spread their terror on Mars. Orographic clouds are also visible, which form when they get higher because of a volcano. The patterns of volcanic eruption and the small amount of water vapour on Mars are also clearly visible.

Patterns of oceans and river channels are also visible, further supported by the evidence from our rovers. Most scientists reason these channel creations with mega flooding in the past. There are some who argue that glaciers carved out these channels. We can also see the impact of craters on dust clouds, where they are clearly blocked, and this must have been prevailing for a long time to be visible from space.

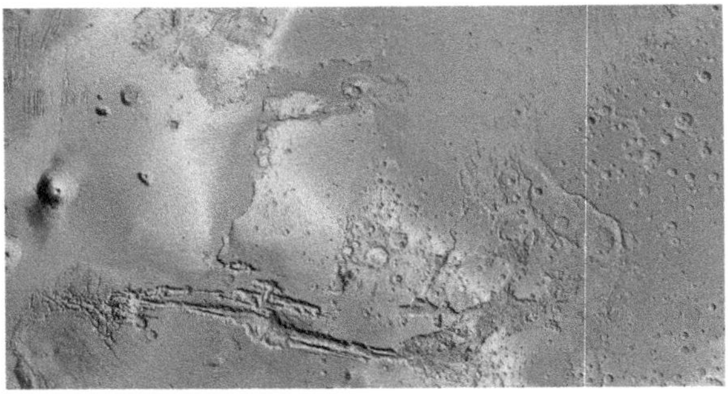

Topographic of Mars(This is not clicked from Mangalyaan but derived upon from its images)

The Standard Ones

- *The Salyut and Mir programs*

When the topic of space agency domination is discussed, especially in the area of space stations, NASA steals the spotlight. They are often attributed to the function and creation of the ISS, and others as just partners. But, the past speaks the truth. The most major contributor to the ISS wasn't NASA; it was Roscosmos/CCCP (two names are mentioned as during the creation of the ISS, the USSR broke up into multiple countries and the major technological systems were still controlled and managed by the Russians).

- While the Americans prioritised the Moon landing, the Soviets had other plans. They believed that before going to the Moon, they should build a base in low Earth orbit. This in turn led to the launch of the first human space station, the Salyut-1. It was an impressive piece of hardware at 18.5 tons, 20 meters in length, 4 meters in diameter, and 99 cubic meters in volume.

This mission is rarely talked about because it is attached to a tragic story. The one and only crew to visit this station didn't live long enough to tell the story. The 3 cosmonauts died during their return flight when the Soyuz capsule de-pressurized, and they were the only astronauts to ever die in space.

- The operation of this station was soon terminated, and it was de-orbited to burn up and be disposed of in the atmosphere. The Soviets tried again, but the second one was lost due to an engine fault in the Proton rocket, and it never reached orbital speeds and fell into the ocean. One year after that, the Soviets launched Salyut-2, which reached orbit but lost altitude control and destabilized, burning up in the atmosphere.

Undeterred, the Soviets launched another space station in 1973, originally meant to be Salyut-2, but since a flight control error prevented it from reaching its intended orbital height, it was renamed Kosmos-557 and presented as a regular satellite launch and not a failed space station deployment.

- The Soviets were finally on a course to success, with the real Salyut-3 achieving orbit in July of 1974 and being successfully visited by the crew of Soyuz-14. This had a literal gun, and the Soviets were the first ones to test fire a conventional weapon in space.

After this mission, the Soviets had an amazing single-module space station series from Salyut-3 to Salyut-7, with no major disasters. However, for the Soviets, this was just the warm-up. In the year 1986, they deployed the world's first modular space station, Mir.

- The United States is able to deploy only a single space station independently till date, and that too a short-lived one, Skylab. The Soviets had mastered the skill of space stations. It took a decade to assemble the seven-module Mir, which would reach a total volume of 350 cubic meters.

Unfortunately or fortunately, the USSR broke during the assembly, marked by the destruction of the Berlin Wall. A new era of space collaboration began between the US and Russia. In the year 1993, NASA started regularly visiting the Mir space station. The Atlantis even delivered the final docking module to the station.

- NASA used these visits to gain information about microgravity and the long-term effects on the human body in space, subjects where they lacked decades behind. The ISS is like a Mir version 2.0 because it's pretty clear where the design approach came from. It is also notable that the first module of the ISS was delivered by the Russians, which provides power and propulsion.

The first crew was also dominated by Russians and launched in a Russian rocket from Russia. The ISS credit is often not equally shared between both, whereas Russia clearly had equal or even more contribution than the Americans.

-The Space Shuttle Challanger Disaster

NASA developed space shuttles to make flights much more frequent to low Earth orbit and rockets much more reusable. It was like an unpowered aircraft. NASA built 6 of these, the first one launching in the year 1981 and the last one in 2011.

- One of the six was Challenger, which got involved in a disaster in its 10th flight and 25th Space Shuttle flight, STS-51-L. It was meant to launch a satellite and check out and research Halley's Comet for 6 days. This flight involved a teacher also, as part of NASA's first and only teacher space program.

The launch was scheduled on a day that would turn out to be a cold and icy one. Even the launch was delayed due to the excess collection of ice over the launch area. At 11:38 a.m., the rocket was confirmed for liftoff, and 73 seconds after, the tragedy struck and took all the passengers down.

- The Americans below celebrating the success were shocked and watched in horror as the shuttle came back down at 333 km/hr of speed. All lives were lost on impact.

Immediate investigation revealed that the 2 rubber O-rings, which sealed the joint between the two lower parts of the right-hand solid rocket booster, were not sealed properly because of the cold and malfunctioned. This caused a breach in the solid rocket booster joint, which allowed burning gas to get into the external fuel tank.

- NASA temporarily paused its programs for the next 32 months, and the investigation committee concluded that NASA was not precautious and was sloppy with things, ignoring its own protocols and engineers' warnings.

This was a major setback, and civilian space travel was halted for a longer period. When it was initiated for the second time, another disaster (Columbia Space Shuttle disaster) struck, killing the last hopes of civilian tourism in space for the next 2–3 decades.

- Although matters since then have improved significantly, with some civilians still travelling to space, it put a permanent spot on the history of space shuttles.

The Present

The Major Ones

-ISS

We have previously covered the events that led up to the ISS. Now, you must have a better context of the innovations that finally caused the invention of the International Space Station. My previous book, From Earth to Mars, also has mentioned the context of the ISS. In this book, I would like to share the modules that make up the ISS and how they are useful.

- Each part was manufactured in different parts of the world and then transported by either of the three rockets: the US's Space Shuttle, Russia's Proton rocket, or Russia's Soyuz rocket. The first piece of the station was the Russian module Zarya. It provides power by its solar arrays and propulsion when there's need to move the station. There are three docking stations in front and one in the back.

The next part was the US's Unity, or Node 1, which was docked with the Russian module using a special PMA (Pressurized Mating Adapter). Unity was attached with 2 PMAs on both sides. It also had 6 docking ports to dock to other components. The Russian Zvezda module was attached next on the back of the Zarya module, which worked as the functional and operational unit of the ROS (Russian Orbital Segment). It also had 3 docking ports in front and 1 in the back.

- Next was the Z1 truss by the US, which held equipment, and was attached on the top of Unity. It was not the main truss but installed for an extra docking port. The next was PMA-3, an extra joint, just in case. The P6 truss was temporarily mounted upon the Z1 truss, which had the first solar array wings to provide the much-needed power to the growing station. Radiator panels were also installed to remove the excess heat.

At this point, permanent human presence on the ISS was made possible, and it has been in continuous occupation ever since. Destiny was installed by rearranging PMA-2. It is the place where most of the experiments are conducted. External Stowage Platform 1 was installed to store spare parts. Canada installed a robotic arm to the station, giving it higher operational feasibility.

- Quest Joint Airlock was installed to allow astronauts for EVA, popularly known as spacewalk. Russia also installed a similar system, Pirs-1, which acted both as an EVA opening and a docking port for rockets.

The solar panel-type structure we see today in the ISS was now being built. It is called the Integrated Truss Structure. S0 truss was attached to the top of Destiny. The MBS (Mobile Base System) built by Canada was added to the side, which could move along the S0 truss. S1 and P1 truss were added on both the sides of the S0 truss. Each side had space for 3 radiators, out of which the centre ones were installed.

- ESP-2 was added right next to the Quest Airlock. The P3/P4 truss was added on the side to add extra solar arrays and radiator panels. Finally, a tiny P5 truss was added on the same side to finish. The following year, S3/S4 truss and S5 truss were added to the other side to balance it out. ESP-3 was added below the S3/S4 truss. Finally, the P6 truss was moved below to the side of the P5 truss.

The Harmony module was added to the Destiny module's side with 6 extra docking ports. ESA's Columbus was attached to the Harmony module. Another robot, Dextre, was added by Canada. The Japanese then added the largest module in the station, Kibo.

- Finally, the last truss was added, the S6 truss. The Japanese then added an Exposed Facility to explore outer space. Poisk was added to the Russian module, which was similar to Pirs. ELC-1 was installed below the truss to store hardware so the station works accordingly. ELC-2 was added to the top of the truss.

Tranquility is added to the side of Unity. A small room is added under Tranquility, the Cupola. The iconic space view of Earth, about which you might have seen multiple videos, is shot from here.

- Another Russian module, Rassvet, was installed by the Russians for storage and had another docking port. Leonardo was installed on the side of Tranquility to store supplies and trash until it can be removed. ELC-3 and 4 were installed soon after.

The Alpha Magnetic Spectrometer was added as a science experiment to test and understand the particles in space. BEAM is an inflatable place which is small during takeoff but expands when it reaches orbit. It was also added as an experiment to test this technology.

- NASA and ESA have planned to reuse parts from it, and then the remaining parts to be re-entered into the atmosphere to burn up and be disposed of in the year 2030.

-James Webb Space Telescope

The James Webb Space Telescope, launched in the year 2022, was launched to investigate our universe's past. How was it formed? How did planets develop? And many other thoughts that fill the mind of a scientist every day.

- Imagine you are 65 million light years away from Earth, somewhere in the Virgo Cluster. You have a very powerful telescope you can use to study Earth. Using the telescope, you would be able to see the dinosaurs on Earth. We are talking about a very powerful telescope. The James Webb Telescope is like a 10 billion USD time machine. It's 100 times more powerful than Hubble, but notably lighter.

It uses the infrared spectrum, whose radiation can penetrate dust clouds and allows us to see the stars forming within. However, as light travels further, it gets stretched and very thin, almost invisible. The James Webb Telescope uses a number of mirrors to reflect the light and rectify it to make it visible to the lens of the telescope.

- The mirror is made of 18 hexagonal segments, each one 1.32 meters in diameter. It allows them to fold the mirror on Earth and unfold it in space. The light is first collected in the primary mirror and passed onto the secondary one. After it is filtered, it is then focused on the highly sensitive infrared detectors, in which photons are converted to electric voltage.

The actual telescope is equipped with 4 tools. NIRCam is the telescope's primary imager in the near-infrared range. 10 sensitive detectors allow it to detect the light of the first stars and galaxies. It is equipped with a coronagraph. Imagine you are standing in front of the Sun. You block the Sun's light with your hand to get a view of the road ahead. This works similarly. It blocks the star to capture objects around it more clearly.

- NIRSpec uses the same spectrum and identifies the patterns to accurately find the chemical composition, temperature, and other things of a planet. But identifying such galaxies takes 100 hours or more on a single one. So, they equipped it with a microshutter system. This allows the telescope to work on multiple targets at once. It is made up of 250,000 shutters which open and close. The shutter system blocks irrelevant light and focuses on what is needed. It also has a constant movement system to always point the telescope in the right place.

For thicker clouds that cannot be captured by the infrared, they have a last tool called MIRI. MIRI can penetrate through thicker clouds also. The problem with it is that it needs to be maintained at a temperature of -266.5 degree Celsius, or it will start capturing its own heat. It has a helium-based cooling system called a cryocooler. To defend itself from our own Sun's heat, it has a sunshield made of Kapton.

- **The Sombrero Galaxy:** Unlike visible light images, which show only the galaxy's bright core, Webb's infrared instrument unveils a smooth inner disc and intricate clumps within the outer rings.
- **Hungry Hungry Black Hole:** Scientists have found an exceptionally hungry black hole at the centre of a dwarf galaxy from the early universe, dating back approximately 1.5 billion years after the Big Bang. We discovered that the black hole was pulling in matter 40 times faster than what we thought was the limit. This intense accumulation, likely short-lived, explains the rapid growth of supermassive black holes in the early universe.

-Perseverance Rover

In the previous chapter, I covered the journey of the rover. However, the components of the rover will be explored in this section. The main purpose of the rover is to find signs of life. It was landed in Jezero Crater, which is thought to have been filled with water years ago.

- The main body of the rover is called WEB (Warm Electronics Box). We have 3 wheels on each side made of aluminium; the front and back wheels can move, and the rover can spin using them 360 degrees. It even has suspension for the harsh terrain. It has a top speed of 0.1 km/hr.

A remote sensing mast is on the top of it for the main camera and sensors, and is the height of a human. The idea is to give a perspective of a human from Mars. The camera on top is the SuperCam, which can identify the chemical composition of objects as far as 7 meters away.

- Below the large camera are the main cameras (Mastcam-Z), which are smaller but can take higher resolution photos and even videos. Navcams are used to help the rover drive safely. Below is an antenna-like structure called MEDA (Mars Environmental Dynamics Analyzer), which studies the atmosphere and environment on Mars.

The rover has 2 microphones, one on the mast and the other on the main body near the wheels. We have Hazcams near the front wheels and back ones to identify any hazards beforehand. It is also powered by a Radioisotope Thermal Generator (RTG), like Curiosity. After solar panel failure in the Opportunity rover due to dust storms, these were implemented.

- RIMFAX is used to study the ground underneath up to 9 meters deep without any digging. A MOXIE machine was installed, which converts carbon dioxide to oxygen, helping humans test it for future missions.

The robotic arm in front has many scientific instruments to study Mars. PIXL studies the composition of Mars on a very small scale. WATSON is another powerful camera to study the rocks in detail. SHERLOC also provides fine imaging. There is a coring drill which collects samples of the Martian surface for future missions to collect.

- As mentioned previously with Curiosity, the rover uses the Deep Space Network to communicate. The rover has three antennas: the ultra high frequency antenna to send high amounts of data through satellites to Earth, a high gain antenna to communicate directly to and from Earth, and a low gain antenna to mostly receive data from Earth.

Important

-Juno

In August of 2011, an ambitious project was sent into the outer solar system, which is an interesting area of research. As you might have already predicted by the mission's name, it was to the planet Jupiter. Juno was equipped with a payload of advanced scientific instruments to study the planet's solid core, map the strong magnetic field, understand the atmosphere and weather patterns, and experience the magnificent aurora.

- It was built by Lockheed Martin and NASA at the Jet Propulsion Laboratory. It was made to withstand the harsh conditions of Jupiter, with very low sunlight, extreme temperatures, and the dangerous radiation of the gas giant. But before it could begin its exploration, it needed to reach there.

It passed the orbit of Mars to perform two deep space maneuvers, which brought the spacecraft back to Earth for a much-needed gravity assist. Juno passed Earth in October 2013, which provided it a boost of more than 8,800 mph, or more than 16,162 km/hr. The instruments were tested and calibrated during the flyby for more reliability, and some beautiful pictures of our planet were also captured.

- Now travelling at 149,000 km/hr relative to the Sun, it was put on safe mode, and all non-essential systems were turned off to save resources as the craft approached Jupiter for the next 2 years. In January of 2016, it began its final approach towards the orbit of Jupiter.

The spacecraft reached the orbit at the highest speed ever achieved by a man-made object, at 208,000 km/hr, and started slowing down. In August 2016, Juno completed the first orbital flyby, which is known as a perijove. It was able to capture many images of Jupiter, showing its bizarre storms and cloud patterns.

- Juno was also able to capture Jupiter's poles in infrared light, which are normally invisible to human eyes. The aurora visible in the pictures is yet to be proven by scientists, although Juno hinted that they may have been caused due to internal processes inside the gas giant.

Juno's elliptical orbit allowed it to explore Jupiter from multiple angles, sometimes even 5,000 km away from the atmosphere, which allowed us to study the intricate patterns formed by the clouds of Jupiter. Instead of a hard rocky surface, Jupiter is covered with a layer of clouds mainly made up of ammonium ices.

- The darker material in these images are thick layers of clouds, whereas the lighter coloured material is clouds higher up in the atmosphere. The gases are said to be the reason behind the unique colors, which include helium and hydrogen. Many images, like those of cyclones, are captured which show different patterns, like a massive cyclone surrounded by 8 other smaller cyclones.

-Chandrayaan-3

With just 14 days of life, Chandrayaan-3 was a wonderful mission which completed more than its objectives. Not only was it India's first successful mission to the Moon, but it cost very little. It cost 75 million USD, less than half the cost compared to China's 172 million USD project.

- In recent years, space agencies have been rushing to reach the lunar south pole because it is predicted that the area has ice just below the surface, which could be used as drinking water and rocket fuel. The Moon is already hostile, but the poles are even worse. In fact, the coldest temperature in our solar system was recorded in Hermite Crater near the lunar north pole.

Although Chandrayaan didn't land in the pole itself, but 600 km away, it still had to experience extreme conditions. As soon as the lander touched down on the surface, the rover began its work. It immediately rotated for the solar arrays to face the Sun. The solar array generates just 50 watts of power, enough to move the rover at 1 cm per second.

- The rover is disabled from sending direct messages to Earth, so it sends its message to the lander, which transmits it to the orbiter, which then finally sends it to Earth. The rover is autonomous, and the team in India decides the point it needs to reach, after which the rover navigates the safest path.

It has two navigation cameras in the front, which helped it in the first few days itself when it was headed straight into a 4 meter deep crater, when the rover turned and changed its direction. The objects needed to survive -200 degree Celsius with the battery to see the next lunar day, 14 Earth days away.

- The rover has a small 10 amp/hour battery, which is not enough to last the whole 14 days. Without power, it has no means of running a heater. India hadn't prepared the rover for the lunar night due to the tight budget. China's rover had a Radioisotope Heater Unit, which used Plutonium-238, a highly radioactive and expensive element to generate heat. ISRO's tight budget couldn't allow the element.

Nevertheless, the 14 days on the lunar surface were a massive success. The rover used a laser to convert the surface into plasma and study it to accurately predict elements on the lunar surface. Oxygen, sulfur, iron, calcium, and many other elements were found on it, which could help us develop bricks suitable for the Moon, potentially useful for future habitation. These are very useful elements that could be mined from the surface of the Moon if humans could.

- However, the most useful experiment was still left. We don't have enough information about the formation of the Moon, and our information is largely theoretical. We believe it has a solid iron core, but there is much we need to study about it.

Chandrayaan drilled a temperature sensor 10 cm deep. Scientists thought of the surface to be cool, around 20 degrees, but it turned out to be 60 degrees Celsius. Just 8 centimeters deep, we discovered it to be -10 degrees Celsius, a 70 degree difference. On Earth, the temperature would barely change.

- Being able to perform experiments on the surface of the Moon is extremely valuable, and only India and China have been able to perform them in this century. In the end, they were put to sleep mode to provide them the largest chance to survive the lunar night.

At last, the lander even performed a liftoff of 40 cm and touched down again to test the lander's activeness.

-New Horizons

In the year 2015, there was very high excitement among the space community as the highest resolution image of Pluto we had was just a color variation and nowhere near clear. However, in the year 2015, this changed. New Horizons performed a flyby of Pluto, the dwarf planet.

- Pluto was considered a not very "interesting" celestial body to explore, but because of the extended efforts by the New Horizons team, the mission was barely approved. In the year 2006, New Horizons was launched under NASA's New Frontiers program for medium-budget space missions. The goal was to get to Pluto as soon as possible.

The launch was one of the fastest back then. It was a light spacecraft on the fast rocket Atlas V. It whizzed past the Moon in 9 hours, 10 times faster than Apollo. On its path to Pluto, it used Jupiter as a gravity assist to shave off 3 years of its arrival time. It also used the planet as a test run and clicked amazing pictures of the planet and its moons.

- After the successful trial, it was put into hibernation to avoid wear and tear during the long journey. During its approach in the year 2015, the scientists turned the systems back on, and the spacecraft sent amazing images every day as it approached the planet.

The images became clearer as the craft approached, and a heart shape could also be identified. Eventually, we achieved an amazing view of the planet. What it sent back was amazing: a planet with an atmosphere, icebergs, glaciers, and detailed photos of the moons of Pluto.

- We found that the planet, considered dead, was geologically active, and it was believed that it has a small heat source in the middle. The New Horizons probe then went on to explore the Kuiper Belt and the objects within, and is expected to now exit the area by 2028–29.

It is in the group of the 5 spacecraft on the trajectory to completely leave the solar system. Its nuclear power is expected to run till the late 2030s, and we can explore interstellar space even further and more clearly with better technology than that of Voyager.

Image Referenced from NASA Science

Standard

-Ingenuity Helicopter

The Ingenuity Helicopter was carried along with Perseverance to test such objects and explore further into Mars, even when RTG power turns off on Perseverance. The helicopter was kept in a tiny black box underneath the rover, and two months after landing, the helicopter was deployed.

- The process took a week to complete. First, the cover is removed and the helicopter is unlatched, carefully folded down, the legs come down, and the helicopter is dropped on the surface of Mars. Perseverance then moves away to allow the helicopter to absorb sunlight.

The helicopter has two rotors that rotate in opposite directions. On Earth, helicopter rotors spin at only a few hundred rpm, but to make flight possible in the thin atmosphere of Mars, the rotors have to spin at 2400 RPM.

- The first flight took place 2 months after landing on Mars, the first helicopter to fly on another planet, and it turned out to be a huge success. It completed 72 flights successfully and was declared dead on January 25, 2024, after a hard landing damaged one or more of its rotors.

Originally designed for only 5 test flights, it operated for 3 years. It flew 30 times further than planned and expected. However, the massive data provided by it is still being taken by Perseverance before complete separation.

- The first aircraft on another planet, that too so successful, was a massive achievement and could define our means of transport in future human colonies without building excess infrastructure.

-OSIRIS-REx

OSIRIS-REx was the first of its kind. The idea of extracting resources has been haunting us for years, and it was certainly a step towards it. It targeted the asteroid Bennu and planned to bring back samples to Earth.

- The rocket was launched in 2016 and reached its targeted location in 2018, and spent two years in orbit around it to study the surface. Finally, it headed towards its primary objective on October 20, 2020, into a crater, dodging obstacles in its way. It plunged its arm inside and excavated 6 tons of debris, out of which 250 grams was collected, and the sample return capsule was closed.

It did its farewell to Bennu in May of 2021 and started its 1.2 billion mile cruise to Earth. After a 2 year 4 month journey from Bennu, the craft approached the place from which it originated.

- On September 24, the spacecraft approached to nearly 63,000 miles from Earth. It powered up and released the sample return capsule at 4:42 a.m. The capsule was angled and jettisoned at the perfect time to land at the exact location, an area of 250 square miles in Utah's West Desert.

Once the capsule was away from the craft, OSIRIS-REx fired its thrusters to avoid colliding with Earth. At 8:42 a.m., the capsule streaked into the atmosphere at a blistering 27,000 mph. The capsule was seen across West America, glowing from the extreme heat.

- The atmosphere rapidly decelerated the capsule, and the samples were subjected to 32 Gs. 2 minutes after entry, it slowed down to Mach 1.4, and the drogue parachute was released, which stabilized its descent. The capsule entered special use airspace at 8:46 a.m., about 10 miles above the Utah defense and military testing range.

The main parachutes then deployed, and it slowed down to 11 miles/hr before touching down on the desert soil at 8:55 a.m. The samples were transported to the Johnson Space Center in Houston.

- The capsule was opened in the Astromaterials Acquisition and Curation Facility, and the samples were studied, curated, and distributed, and will be for the years to come.

The journey of the spacecraft is not yet over. It was put in an elliptical orbit around the Sun, crossing Venus's orbit on multiple occasions. In 2029, it will chase and track down Apophis, a 1200-foot stony asteroid, destined to make an exceptionally close flyby to Earth, and the journey will continue.

Future

Major Ones

-The Artemis Program

The Artemis program is a step towards a massive change. During the 1960s, planetary human missions were promoted. However, as soon as the Apollo missions ended and the US won the space race, the motive was diverted. Exploring other planets with satellites was promoted, and space stations were being built. Reusable rockets and shuttles were being developed, and human missions were limited to low Earth orbit and the ISS.

- Now, NASA has again changed its motive for the second time. The missions to low Earth orbit and other non-essential missions that include satellites are transferred to the commercial industry, which does the job cheaper and with more reliability. NASA has shifted its focus to developing a colony on the Moon and Mars.

The Artemis program targets building a colony on the Moon. If everything goes according to plan, we will have a colony on the Moon by 2034, sooner than you have ever imagined. In fact, the first modules of the colony will be assembled in this current decade. The next step after this will be Mars. Artemis 1 has already completed its journey.

- Imagine you are on your personal spacecraft. You go to the Moon and dock at its private space station, you board upon a lander from which you deboard into a rover, and when the taxi service ends and you open the door, you find yourself in a base on the Moon.

The initial layout of the colony, as planned by NASA, is as follows: first of all, your personal spacecraft docks on an ISS of the Moon, the Lunar Gateway. The Gateway has a lunar lander attached, which will take you to the primary landing zone on the lunar surface. Near the area, a pressurized lunar rover will be ready to carry you to the 500–600 meters away lunar colony.

- From there, a pressurized tunnel will connect to your rover. As soon as you enter, you will find yourself inside the habitat module of the colony. You can keep your spacesuit there and move forward. You will see a living room, which will have all the life support and controls.

For food and luggage, the living room will connect to the logistics area, where things can be kept. If you move further, you will come across the food production area, where food will be grown. A solar park will be set up nearby the colony to provide sufficient power. A propellant production and transport machine will be set up nearby to produce fuel for the rockets.

- The mission has already begun and will likely continue, so this isn't just imagination. After the Artemis mission starts, we can expect astronauts on the Moon every 6 months. Space tourism to the Moon won't be a faraway reality either. It might be that you can go to the Moon someday. Artemis 1 has already launched and succeeded. Artemis 2 mission is also under process while writing this book. If Artemis is successful in establishing its base, Mars isn't much far.

The plan of each mission is mentioned below—

Phase 1: Build, Test, Fly

- **Artemis 1:** Although a test flight to the Moon, it performed many maneuvers, and the fake astronaut helped us understand human pressure. This flight is already completed and was a massive success, giving large hopes for the mission and gaining support from the public.
- **Artemis 2:** This mission is repeatedly delayed due to launch window errors and is expected to launch in April with 4 astronauts, including the first female one. It will be a flyby of the Moon and the real test of the Orion space capsule and its life support systems, as it is a new technology developed by NASA. It will travel further than any human ever has into deep space.
- **Artemis 3 (Targeting Mid-2027):** Originally planned as a landing mission on the surface of the Moon, the flight will now act as a demonstration of the docking module between the Orion spacecraft and commercial lunar landers, and evaluate the new Axiom spacesuits.
- **Artemis 4 (Targeting early 2028):** The mission is now planned to land humans on the surface of the Moon after 6 decades. 2 astronauts will stay on the surface of the lunar south pole for a week while the rest of the crew orbits above.

Phase 2: Early Infrastructure (2028–2030)

- **Artemis V (Late 2028):** Plans to launch the second lunar landing and the start of building the Artemis Base Camp, and delivery of the first unpressurized lunar rover.

This phase will mark the start of an iconic space era where humans travel to the Moon every 6 months.

Phase 3: Permanent Presence

- **Artemis VI (2031):** Delivery of a crew and science airlock and the third crewed mission.
- **Artemis VII (2032):** Deployment of a habitable mobility platform (a pressurized lunar cruiser) to allow for 30 days of surface stays.
- **Artemis VIII (2033):** Delivery of foundational surface habitat, extending the stay to 60 days.
- **Artemis IX (2034):** First use of the upgraded SLS Block 2 rocket to deliver heavy logistics for expanding the base.
- **Artemis X (2035):** Focused on establishing long-term habitation, with stays lasting up to 180 days.

From here on, missions will last 6 months each like the ISS, and Mars will be the next step.

-Mission to Mars

NASA has planned to explore Mars after the Artemis mission and follow the same tested path. SpaceX has a faster and more worked-on plan right now, executing this year itself, hopefully. If we are successful in completing this plan, we will become a multi-planetary species.

- Though SpaceX is determined on its mission and has high ambitions, this mission requires more than heavy rockets. Unlike other nearby colonies like the Moon, which has a travel time of just a single day, meaning everything is available at their footsteps almost immediately, a Mars mission would require self-sustainability.

Travel time to Mars is 8 months, and even if we develop nuclear propulsion, a 45-day travel time would still be dangerous. The fastest communication will take 11 minutes, which means no quick response to and fro. They have to harvest everything that the planet has to offer, and to be honest, it doesn't have much. However, we will create massive industries that will convert those resources into necessities.

- The plan is to habitate the planet with robots, not a group of 4–5 but hundreds. These robots will perform all functions that the humans will perform after they arrive. They will test all the systems and develop the Martian colony. Yes, Mars will be captured by the robot army before humans.

Optimus, the Tesla Bot, is being manufactured for the exact purpose. SpaceX has planned to launch the robots in late 2026, with the secondary launch window in 2028. 5 Starships will carry 10 tons of payload each to Mars. The real reason behind the 5 rockets simultaneously is that hopefully one of them lands safely. It is expected that we have a learning curve, and SpaceX specifically learns from its mistakes to master the tech.

If we assume that one of them lands safely and the handful of robots are unloaded, they probably won't have much work. 2 years after the first successful landing is when the real colonial fleet arrives, with 20 Starship landers with each 75 tons of payload. This payload is going to carry mostly infrastructure for the first people to arrive.

- The Arcadia region on Mars has been identified as the primary settlement area. It is just east of the solar system's tallest volcano and is believed to be a flat plain and has large deposits of water ice. Water is the biggest challenge on Mars.

The robots shall be building two main machines on the surface of Mars for their human comrades. First is the water treatment plant. Bringing water from Earth will be very heavy and will possibly be the biggest challenge. Water is a very useful resource and perhaps the most useful one on Mars. Water isn't just needed for drinking; it is made up of oxygen and hydrogen, which can be broken down for future Mars inhabitants to breathe. Hydrogen is an important component for propellant (rocket fuel).

- Going to visit the Red Planet will be a one-way ticket for the next century at least. However, the first colonists will need to come back until there's a proper colony built there. Therefore, the second machine will be a propellant recycling and generation plant.

Hydrogen is the lightest element in the world and is therefore hard to control. SpaceX uses methane as fuel for its Starship, which comprises 1 carbon for every 4 hydrogen. The carbon stabilizes the hydrogen just enough. Fortunately, we also have carbon on Mars, as the thin atmosphere is mostly made up of carbon dioxide. You get carbon for the fuel and some extra oxygen.

- Now, to support the production of such technologies and water, mining will have to be set up. A mining rig machine will be set up, and it is hoped that water ice is found. The mined surface can be used as raw material for roads, landing pads, and other land infrastructure.

This is where we would eventually end up after our first large-scale robot landing. 100 landers with 150 tons of payload each will come to colonise Mars as soon as these steps are proceeded towards.

- The next thing to be established is a power plant. After humans arrive, power needs will rise exponentially. After the power plant is completed, all the systems will be activated that the rovers developed. The life support systems that provide oxygen, heat, and air pressure will need to be activated 24/7.
- The length of a day on Mars is fairly similar to Earth, meaning the sleep cycles might not be disturbed as much. Every 12 hours, it will be nighttime. Unlike Earth, where places powered by solar panels are supported by the grid at night, Mars is completely different. Power means life.

Large batteries are being developed by SpaceX to provide for this need. The landers can become shelter for a few days at least on Mars. An ice shelter is being thought of on Mars, as it keeps heat inside and acts as a layer between the intense radiation. The idea is to make an inflatable dome with a layer of wall, finally insulated with ice.

- We can cover it with local materials of Mars if we need a more solid structure for research. We need something like concrete to hold such a structure, so a binding agent will be needed to use the local materials. Luckily, we have a concept known as biocrete or astrocrete. Human blood and urine act as very good binding agents, turning the material as hard as concrete. Urea can make the material stronger than what we use on Earth today.

The next issue is with food. Mars soil is toxic. It has perchlorates, which have chlorine, which is mixed into swimming pools to avoid growth of organisms. Another facility will be needed to filter out the soil and enrich it with nutrients, possibly waste. This isn't going to be pretty, but possible.

The Important Ones

-Europa Clipper

The single most important thing we know of on Europa is water. Just imagine all the water on Earth, and double that. That's the amount of water we believe to be on Europa.

- To explore the reality further, NASA launched Europa Clipper in 2024. It is scheduled to reach there in July of 2030, faster than Europe's JUICE, which was launched earlier. Europa sits at one of the most dangerous points of the radiation environment around Jupiter.

Europa, the moon of Jupiter, is expected to hide a massive subsurface ocean underneath its icy surface. It is believed that Europa has the necessary elements to support life: water, energy, and chemical building blocks. It is the first time we are dedicating a satellite to this moon of Jupiter, which will study the water and ice underneath.

- It's not a life-search mission, rather a potential habitability one. The Clipper has 9 instruments, out of which 5 are called remote sensing instruments, as they study the moon using the reflection of light. The other 4 study the environment around it, like a sniffing device from senses like smell.

It is the largest one made by NASA ever for a planetary mission, and weighs around 6.5 tons. Its solar arrays are very large to support Jupiter's distance from Earth. Jupiter receives only 4% of the light from the Sun than Earth receives. They are big and powerful, as they have to withstand the intense radiation from Jupiter for the rest of its life.

- Jupiter acts as a large particle accelerator, and electrically charged particles are trapped in its magnetosphere. These particles slam against Europa, and the craft will face them too. NASA uses two ways to protect the craft from severe damage.

Instead of achieving an orbit around Europa, the orbit is set around Jupiter, and the craft flies by Europa multiple times. A radiation protection vault is also inbuilt inside the spacecraft, which carries all the computers and sensitive electronics that could be influenced by the radiation.

- In each flyby, the craft sees the equivalent of a million chest X-rays. The gap between Earth and Jupiter isn't too difficult for NASA. It will complete a flyby of Mars and then of Earth to use gravity assist and speed the spacecraft up.

The gravity assist from Mars was completed in 2025, and the Earth one is in December of 2026. This will be a really exciting moment for all of humanity, because if we find evidence of water, the most essential resource probably, it could change a lot of assumptions. It is a golden opportunity to explore something we don't know much about.

-*Gaganyaan*

Under the leadership of ex-ISRO chief Dr. K. Sivan, he dreamed of something similar to the US's Apollo and Russia's Vostok, a human space program for India. Under the political regime of the BJP, this was heavily pushed, and the honourable Prime Minister announced the project on Independence Day in the year 2018.

- This is a massive commitment, considering that only 3 space agencies have been able to achieve this goal. The mission is significantly delayed as of the current status. The final goal of the mission is to transport Indian astronauts for the first time to low Earth orbit (400 km) for a duration of 3 days and land them safely in the Indian Ocean.

The trials have significantly moved forward, and many have been completed. A Vyomitra robot is planned to be sent to develop basic understanding, followed by 2 or 3 uncrewed trial missions before launching the real one with astronauts in early 2027.

- Astronauts were trained under and by Russia, a country with much prior experience, although the pandemic halted the progress significantly. This is a major leap, as humans in space hold a very high risk, and developing the technologies is often followed by a learning curve.

In the past, the world was never this interconnected, and a failure could be hidden with just a few steps, but today news spreads like fire, budgets are cut, and people are fired. The mission holds very high stakes for the Indian space program, because if this fails, ISRO may not get the budget for this for the next decade.

- The mission is almost ready, and 8,000 tests have been successfully conducted. ISRO is ensuring that not even a single number is missed. Building a new capsule is a complex technology, and the past represents something dark and dreadful.

-Dragonfly

Saturn's moon, Titan, keeps one of the largest truths of the universe a well-enclosed secret. While life on Earth erased its own birth, Titan is said to be very similar to Earth before the birth of life. This gives us hope of understanding how life was born initially.

- The moon is the only one in our solar system to have an atmosphere which is even thicker than Earth, in fact 4 times more. The atmospheric composition is also somewhat similar, with the majority composed of nitrogen. However, instead of oxygen, it has a methane-based weather system, which we call natural gas on Earth.

The moon faces multiple issues, and its greenhouse effect acts in reverse compared to Earth. The gases deflect the little sunlight the moon gets, and the methane falls like liquid on the ice, which is frozen as hard as grained rock. The result is flowing rivers, seas, and oceans of methane, making it the only celestial body apart from Earth to have a flowing hydrologic system with rivers.

- It has deserts and even dunes made of some organic matter. When we put all of these discoveries together, we find something similar to ours.

We visited the moon previously with Huygens, as talked about in this chapter, but it didn't last very long. So, NASA is sending another one. Dragonfly is different. It's an octocopter (helicopter with 8 rotors) and can explore the surface by moving itself around and exploring different aspects of the surface.

- It will have all the instruments we need to measure and finally end the debate, hopefully. It will be a super cool mission and is yet to launch. It will use the tested method of electricity generation in places where sunlight is scarce. A Radioisotope Thermal Generator (RTG) will power the rover using a nuclear compound's heat.

It will charge itself during the 192 hours of nighttime on Titan and then fly up to explore Titan during the day and communicate with Earth. It has a range of 72 km, which is a large number, as even the Curiosity rover on Mars has traveled 21 km in 7 years.

- It will present us a world, hopefully similar to Earth, and help us trace how we were developed and what is the reason for life.

-Jupiter Icy Moon Explorer(JUICE)

The European Space Agency is the first one after the United States to send a satellite towards the outer solar system. Launched in 2023, Jupiter Icy Moon Explorer will overlap at some point with the Europa Clipper in 2031, when it is scheduled to reach Jupiter and explore its moons.

- Unlike Europa Clipper, the satellite will focus on all the moons of Jupiter rather than a massive focus on one. The mission will study Europa, Ganymede, and Callisto, and the 12-year project will end with a dedicated orbit around Ganymede.

The craft will use multiple flybys, and became the first one to get the gravity assist of both Earth and the Moon in August 2024. It then completed a gravity assist of Venus in August of 2025, and is expected to fly by Earth in September of 2026, then finally complete its last flyby of Earth in January of 2029.

- The mission will use multiple flybys to gain very high speeds, but with a longer route than Clipper. The satellite will then reach Jupiter and complete 35 flybys of its moons. It will be the first satellite to capture an orbit around Ganymede and is designated to study water and life on Europa and Callisto.

It will be a dangerous journey, as using gravity assist in the right way is a hard-to-achieve technology, and derailment is a haunting reality. Being the first one after NASA to do this isn't easy, and requires lots of resources and money to develop such technologies.

- Even a single error could flop the whole mission, but it is on the planned route right now. We all hope that it completes its journey, as intelligence from our previous satellites, the Clipper and JUICE, could help us all end up on the same conclusion.

Standard

-Tianwen-3

China aims to return what NASA dreamed of for years. The Martian soil is precious and a necessity to confirm information and build a colony with trust rather than failure.

- China will send two spacecraft to return with samples of the Martian soil. A lander/ascent module will land on the Martian surface, and an orbiter/Earth returner will bring back the sample from the orbit of Mars to China. The two crafts will be launched separately and will offer higher efficiency, fuel, and payload capacity.

The mission is planned in the 2028 Mars-ejection window and can help us develop our strategy for the Martian surface. With something solid about Mars rather than just photos and radio waves from our rovers, we can confirm more about the theory of life on Mars.

- It will also be the first spacecraft to fetch samples and bring them back to Earth from another planet. This is completely different from NASA's approach to the asteroid, which completed all its objectives in a single spacecraft and reused the craft for other research, offering higher sustainability.

However, Mars is a complex place to land, and the communications are very slow. This will provide solid evidence upon the discoveries of our rovers, whether from China or NASA.

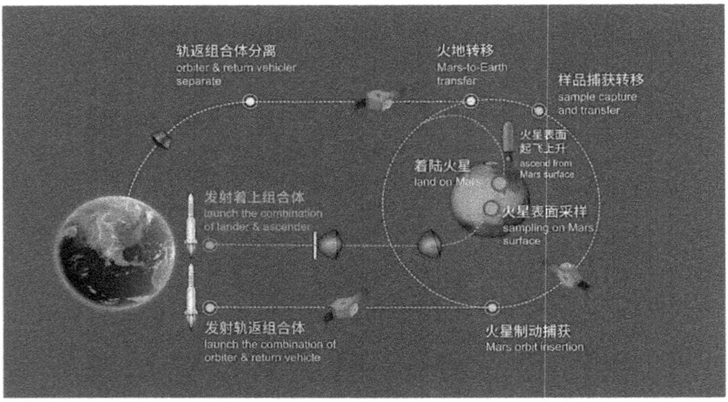

--

-LUPEX(Lunar Polar Exploration Mission/ Chandrayaan-5)

LUPEX is a joint collaboration among ISRO and JAXA (Japan's space agency) to send a lunar lander to the exact pole of the Moon. Chandrayaan 3 did land near the south pole, but not on it. This is not because ISRO has mastered the landing technique (which is a cherry on top); JAXA had planned this years before ISRO's success.

- NASA and ESA have also provided equipment to use in the lander and rover, and because others are collaborating, it has a nice budget. It will be able to survive the lunar night, because this mission will have an RTG and ISRO can finally purchase plutonium, which China used.

It will be launched after ISRO's Chandrayaan 4, which will fetch and bring lunar surface samples to ISRO. The launch vehicle will be the one from Japan and will be manufactured by Mitsubishi.

- ISRO re-engineered the lander for some specific needs, and the rover will now be mounted externally rather than kept inside the lander like Chandrayaan 3. If the Artemis program works accordingly, and the first parts of the colony are successfully deployed, this rover may be in coordination with the mission, and the colony could be captured and found by the rover.

The rover and lander have the most latest and advanced equipment, and are likely to help us set up a colony on the Moon.

India's Future Space Missions Roadmap

Mission	Description	Target Timeline	Significance
	Moon rock sample return mission	October 2027	First Indian lunar sample return
	Replacement satellites for NavIC navigation system	By September 2027	Strengthens India's GPS alternative
	Venus Orbiter Mission India's first mission to Venus	March 2028	Expands interplanetary exploration
	BAS-01 First module of Bharatiya Antariksha Station	2028	Beginning of India's space station
	Chandrayaan-5 (LUPEX) Joint India—Japan lunar south pole mission	September 2028	Advanced lunar exploration collaboration
	LVM3 Upgrade Semi-cryogenic upgraded heavy rocket	2028-29	Higher payload capacity & efficiency
	Third Launch Pad New launch pad at Sriharikota	2029-30	Supports next-gen rockets & higher launch frequency
	NGLV (First Flight) Reusable heavy-lift rocket	September 2031	Reduces launch costs significantly
	Bharatiya Antariksha Station (Full) Complete 5-module space station	2035	Independent long-term human presence in space
	Crewed Moon Landing Indian astronauts land on Moon	2040	Historic human spaceflight milestone

India's future missions(Image credit to OpenAi)

-Martian Moon eXploration(MMX)

JAXA is aiming for a range of elite missions. Extracting and studying the Martian soil is a dream come true for many scientists. China's Tianwen-3 plans to get the Martian soil.

- However, as you might know, Mars has two moons: Phobos and Deimos. JAXA plans to get soil samples of the moon Phobos to Earth and observe its patterns. As we are getting closer to the dream of a Martian colony every day, the information about its surroundings and whether they have something to offer is interesting and even more needed.

We cannot return to Earth for resources frequently within a short timeframe, but if we need something urgently, Phobos and Deimos will always be in a 1–2 day transfer window. Setting up a pre-land or refuelling station on the moons is completely possible, considering that they do not have their own atmosphere, and people can prepare for a Mars landing there beforehand with all essentials.

- It can also hide some secrets about the creation of moons, although they are expected to be asteroids that were captured by Mars itself. In space, even the tiniest information paves the way for massive innovations to come in the next century.

It is due to reach Mars by 2027, and will launch in the November 2026 launch window. It is expected to come back to Earth with the samples by the year 2031 and provide us the first samples of the moons of Mars.

-Shukrayaan-1

"Shukra" is the Hindi terminology for the planet Venus. The most successful missions to Venus were sent by the USSR, which also couldn't last much long. However, in the terrible conditions of Venus, scientists still suggest that microbial life could exist 50 km above the surface, where the temperature and pressure are similar to Earth.

- JAXA, NASA, and many other space agencies have their orbiters around Venus, our nearest planet. Therefore, ISRO is sending its own orbiter with more tools to study the environment and atmosphere of Venus, and more importantly, give a testimony of its interplanetary launch skills.

In space, though we have many orbiters around Venus, some facts are kept secret. To get the same amount of intelligence as other countries on the planet, ISRO is planning to send this mission after its flagship programs in the year 2028, and its journey is expected to last 112 days.

- The orbiter will last for approximately 4 years, and ISRO will send a descent probe. The descent probe part has always remained a gamble in Venus's harsh conditions. Only the USSR has been able to soft-land probes on the surface of Venus which actually transmitted signals back, and if ISRO is able to achieve the landing, then it will become the second to achieve this hard task.

This will be a shocker to the whole world and send a message of ISRO's superiority in space probe landing systems. The government has approved the budget, and there are no major obstacles for budget issues or launching systems. The launch dates have also been announced, but the mission faces a delay due to the flagship Gaganyaan mission.

PART III

THE NEXT SPACE AGE

CHAPTER FIVE

HUMAN LIFE IN SPACE

> *"Looking back at Earth from the moon... Everything you've ever known is behind your thumb, and that blue-and-white ball is orbiting a rather normal star..." — Jim Lovell*

The human body adapted to air and gravity, and now on Earth, it feels almost as if it doesn't exist. However, in the endless void of the universe and space, none of these exist.

- Establishing a suitable air pressure is relatively easy, but simulating gravity is a complex subject. Therefore, the spacecraft sent into space generally let astronauts live without gravity. Sure, Mars and other planets might have gravity, but its significant difference from Earth's can only reduce effects, not completely erase them.

Adapting to the no/low gravity environment will need years of constant habitation, and people that will be born on Mars from childhood might be able to adapt till their adulthood. The human body is highly adaptable and will likely adapt within years completely.

- In this chapter, I aim to show you a potential lifestyle in space, Mars, the Moon, and other celestial bodies, and the physical and physiological impacts.

HUMAN LIFE IN SPACE
THE ISS

The ISS represents a feeling of somewhat zero gravity. Although it does have air, so astronauts do not have to wear spacesuits all the time. It is tightly packed with equipment all over. Moving around in such a confined area, especially when in a free fall, can feel difficult to them at first, but control slowly becomes routine using guiderails and training.

- The fluids in your head and ears get squashed, and the mind is in a state of confusion and disoriented. However, astronauts do not come to have leisure time in the station. They follow a strict work schedule every day. Mission control back on Earth, located in multiple places like the US, Europe, Russia, and Japan, rules out their schedule.

The schedule is tightly packed and broken up into chunks of 5 minutes, so the astronauts do not experience boredom. When humans experience boredom, we tend to become more impulsive, tense, irritable, and volatile, which is definitely not required when everyone is bound to a confined space with no way out. They follow a 12-hour work day, as mentioned previously, broken down into chunks of 5 minutes so they don't have time left.

- The ISS follows Greenwich Mean Time (GMT), upon which all countries agreed. The work day starts at 6 a.m. and lasts till 9:30 p.m. Astronauts begin their day like everyone else, getting cleaned up and ready for work. However, here is where confusion grows.

Water acts weird in space and does not go out to the drain but floats around everywhere. It tends to stick together in spherical shapes and float around. Water sticks to other objects naturally. So, to bathe, they use a wet cloth with soap and clean themselves up with it, much like a sponge bath. Toothpaste can't be spit out, so it is swallowed instead.

- The ISS completes an orbit around Earth in 90 minutes, meaning that the astronauts experience 16 sunsets and 16 sunrises every Earth day. That might sound fun at first, but health deteriorates quickly. The astronauts follow a 24-hour Earth day, and lights are used to mimic the Earth day and the natural clock. This helps them defend themselves against orbital insomnia, which plagued previous generations.

You can't bring your wardrobe with you to space. Astronauts typically wear a shirt for 2 weeks. Pants typically get worn for a few months, and when the clothes become unusable, they are thrown in the trash. Vacuum toilets are used, and when all the trash and dirty things are collected, they are ejected and burn up in Earth's atmosphere.

- The ISS uses a closed water system, as a fresh, non-recycled one would be unsustainable for the long term. On average, the ISS uses 70% recycled water. The non-recyclable water is used towards extracting hydrogen and oxygen, and oxygen is pumped into the ISS as air for the astronauts to breathe.

Two types of food are consumed mainly: thermostabilised food or military food, which is cooked at high temperatures to kill pathogens and enzymes, and second is freeze-dried food. The freeze-dried food is normal food with moisture removed. Astronauts simply add hot or cold water to rehydrate the food and eat it.

- Astronauts, however, can't have food of their choice. For example, bread isn't allowed as its crumbs could potentially block the equipment. Soda drinks aren't allowed as without gravity, carbonation doesn't separate in your stomach and just builds up, leading to wet burps or just vomiting.

On every Sunday evening, everyone comes together to celebrate with a ritual pizza party. The calories are definitely needed. Each astronaut needs to put in 2 hours of cardio and resistance training every day. They have treadmills, stationary bikes with straps to hold them, resistance bands for muscles, and a machine which provides a full-body routine using vacuum pistons.

- We are doing this as the body doesn't need to fight gravity in space, so muscles weaken and bones lose density, which could make you sick on the station, but worsens when you arrive on Earth. Even with a workout routine of an athlete, space could still fail you.

Without gravity, your bodily fluids accumulate in your head, making you feel congested and heavy, known as Space Snuffles. A more serious health problem is nearsightedness in space. There is no clear explanation, but it is being linked to the pressure caused to the skull by the change of fluids in the head and spinal cord.

- The ISS is a science lab, so scientists observe how biological beings react with no gravity and such strange conditions. This can be used later to understand aging and growth on Earth and work in environments not possible on Earth.

The Moon

As we come closer to the Moon every day, and with Artemis to build a colony, lots of research is undergoing on human behaviour in the Moon's low gravity. Lots of behavioural, physical, and chemical changes occur in areas outside Earth.

- The Moon is outside of Earth's magnetic protection from the Sun, and therefore faces large amounts of radiation. We could build a Mars city if we are able to transport 1 million tons of cargo to the planet. However, the complexity isn't the ships or capacity, because it's possible, but the transfer window. It would take decades to catch the transfer window again and again and launch the rockets.

On the other hand, the Moon shows a different image. We can possibly build a city on the Moon more easily, as the transfer window is much wider and the distance is shorter. The Moon still presents many challenges as well.

- The Moon has significant temperature swings, from -175 degree Celsius to +126 degree Celsius. A typical Moon day consists of 29.5 Earth days, which is very extreme. The daytime temperatures are the most threatening, as computers do not like heat.

Mars doesn't have this issue; it's mostly cold at 20 degree Celsius to -70 degree Celsius. If we want to legitimately live there, it has to be highly insulated. However, human life on the Moon isn't as hard. We will have to study the long-term effects of habitation and be protected, but unlike Mars, which is a 3-year journey at minimum, the Moon doesn't restrain us.

- The journey to the Moon is a minimum of a day and is expected to be even shorter with nuclear propulsion technology. The Moon could offer a lot of space for habitation and research, with hundreds of millions living there. The continent of Asia, which is similar to the size of the Moon, has 5 billion people.

Oxygen is also not scarce on the Moon, with each cubic metre of regolith (Moon soil) containing 630 kilograms of oxygen. A person needs 800 grams of oxygen every day to stay alive. This means that a cubic metre of regolith could support a person for 2 years.

- It is believed that water is also hidden, waiting for us, which would have been brought from asteroids. Just burning the surface of the Moon should release water vapour, which we could trap and liquify.

NASA has already figured out that food loses nutrients over time, so they will need to grow vegetation in space to sustain life. They are already performing farming experiments on the ISS. They have been able to grow leafy vegetables, which is enough for vitamins.

- Communication is essential, not only for survival but to keep people sane. Direct communication from Earth in real time, live streams, or whatever makes them believe they are still part of the human race is needed. NASA's Deep Space Network and SpaceX's huge satellite network are extremely useful.

We have not figured out much about the effects of the Moon on humans, but it is expected to be similar to the ISS or EVAs. Power will be extracted from solar panels initially, but Helium-3 (from the huge fusion reactor, the Sun) can be used to source electricity later.

The Mars

Mars will break down human endurance the most. It will make us reach our primal stage where survival came from community, which has to be earned. The journey itself to the planet will be cruel, with muscles weakening, eyes losing strength, and fluids disrupting, along with the toll of isolation and confinement. However, the transition upon arrival will be even worse. The transition from Earth's gravity, to microgravity, to Mars' 1/3 of Earth's gravity will disorient their mind further. Astronauts are likely to experience gravity sickness, which causes feelings like heaviness, nausea, and being uncoordinated as muscles struggle to adapt.

- The atmosphere is only 1% of that of Earth and comprises carbon dioxide, offering no breathable air. Mars has no magnetic field, which exposes astronauts to the Sun's direct cosmic rays and solar storms. To sustain life, we will have to establish a closed-loop system, where air, water, and waste are recycled and reused until it completely gets exhausted and finished in a literal manner. Exercise routines would still have to be heavily maintained, and constant communication is needed for the sanity of the person. NASA and many others have suggested ice homes, as they offer significant protection from the Sun. The houses will comprise a mixture of multiple styles, with some inflatable, some made from Mars itself. 3D printing technologies and robots will help in construction a lot.

Communications from Mars take 4 to 24 minutes depending on the orbital location, and are definitely needed. Multiple relay satellites will need to be deployed other than the deep space network to allow for more reliable communication. Health data, emergency signals, and research will also be spread through this. Real-time communication is not possible in this era's technology, so the communications will be a mix of emails and recorded video messages. AI companions and robots will be needed to talk to the astronaut like a friend. Helpful and emotional systems, which can make everyone feel home-like and act as everyday heroes in our life on Mars, will be sent, and the atmosphere is planned to be an active community rather than something boring.

- Hydroponic technicians will help operate Martian farms, taking care of plants, monitoring water systems, and harvesting crops. Cooks will transform these basic ingredients into delicious and nutritious meals, finding workarounds in limited things. You can't eat freeze-dried meals for your entire life. General maintenance workers will clean the habitats, fix the robots, and deal with the constant onslaught of Martian dust. Environmental operators will help us manage the recycling and resource management systems, as such professionals can help us make use of each and every drop of resource, not letting go of anything as waste. They will also ensure that everything is up to date, and that air quality is high.

All of these things that we take for granted on Earth are a question of life and death on Mars. Disposing garbage will also be a massive undertaking and a high level of innovation. Teachers will teach younger settlers both Earth-based knowledge and essential survival tips on Mars. Health and wellness workers will help settlers combat the heavy effects of such environments, forming specialized equipment and routines for specific needs. They will not have Earth's luxuries, but live in a society where everyone's contributions are valued and important.

- It is going to be hard, but the value of hard work is going to be higher on Mars than Earth. With only a few settlers (hundreds or thousands), settlers will experience a level of isolation and interdependence never exposed to most of us. This community will have to foster a commitment to purpose, where each one feels connected to its success. The lack of privacy and inability to escape when in tension can escalate minor issues into major problems. Isolation and the strange atmosphere can lead to severe problems such as cabin fever or sensory deprivation.

Over time, Martians will develop their own culture, out of necessity and ingenuity because of the limited resources and ways to celebrate. They will get creative and find ways to celebrate milestones together, finding new ways of entertainment. They will bond over the sense of isolation and limited resources, and being part of something historic. This will help them develop pride and identity for the colony.

- Healthcare will be unlike Earth, where the different environment will offer unique medical risks, and the risks will be heightened. The curated and confined area in which they would live inside could lead to long-term mental health risks, such as anxiety and depression. Injuries from falls, equipment malfunctions, burns, and many other things could turn life-threatening very quickly. Humans may be able to restrain the stress of Mars for the short term, but it may pose health risks in the long term for which the astronauts should be prepared.

Being old in such situations is also not acceptable, as it is heavily taxing on resources to sustain another life with limited resources. The approach of Martians towards the end of life might be harsh. All astronauts must be trained heavily on basic first-aid equipment until a medical professional is ready to take over. Robotic helpers or telemedicine help from Earth could potentially ease the risk, but the communication delay means that the people on site need to be trained enough to handle the situation themselves, and life largely depends on their skills.

- However, as life adapts and they survive and thrive, the new frontier outpost could turn into a thriving and well-prepared city. All these health risks may slowly start to decline, especially the ones related to isolation and confinement, as cities expand and communities enlarge. One of our largest dreams is terraforming Mars and making it green again, even if slowly. The process would begin gradually, with releasing water to make lakes and planting trees in the regolith in small areas and expanding section by section. This may also lead to the creation of large patches of greenery on the surface of Mars, but this future can only be imagined in the distant future. Introducing microbial life to the surface, and potentially creating zones where you can breathe openly, could gradually create pockets of life on the planet and make it completely habitable one day.

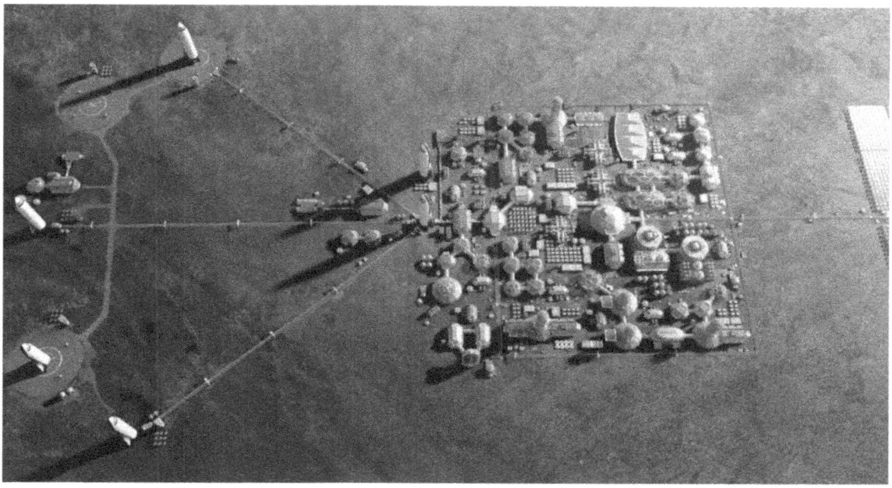

-

The human body will adapt to space very soon. The next generation of settlers, which will be born in space, will exhibit the adaptability of the human body, and soon we won't face as many issues as we face today.

- It is just the beginning of an era, which will change the lives of all generations to come.

The main problems in travelling long distances are communication and gravity. Gravity causes physical and chemical changes in the body, causing many health-related risks, while communication plays a huge psychological role.

- The human body will naturally adapt to gravity, sooner or later, but the psychological effects will only vanish once a decently sized population shifts there, making a decent amount of area and people to talk with.

All issues like isolation, no privacy, loneliness, and other psychological disorders will significantly decrease even if a million people shift there or are eventually born.

CHAPTER SIX

The Future of Space

The future of space covered is the part which we can't access. The part which isn't hidden, but too far away. This includes our solar system within our galaxy.

- You would expect a million-year-old society to achieve this, but we simply can't.

Proxima Centauri is the nearest star to our solar system after our own Sun, only 4 light years away. We haven't set our foot on another star or solar system outside our own, which includes robots as well.

- We are simply too negligent towards these.

In recent years, our interest has risen. We sent Hubble and James Webb telescopes to investigate what we don't know.

- The future also consists of technologies which might be able to make this a reality.

In this chapter, I cover these to provide you a wider view of the far-sighted future of our civilisation and what we have to do to achieve it.

THE SPACE MISSION

PROXIMA CENTAURI

At our current speed, it feels like Proxima Centauri is just a distant dream. Let me explain. One of the closest planets to us is Mars. To reach Mars, we need to travel for 6–8 months. To travel to the outer solar system, it takes us a decade. So, to travel to another solar system, a century-long trip sounds illogical, but at our current state, it's the reality. However, the answer to reach this star faster, and by that I mean way faster, is hidden in our own past.

- For most of human existence, we travelled through boats. The boats were driven by the force of sticks used by humans, or cloth sails driven by the wind. The sailing ships never carried fuel but crossed vast oceans, harnessing the power of nature. Now, what would happen if we applied the same principle to space?

Well, first let's understand how the boat worked, because it might reveal something surprising about propulsion beyond Earth. If I were to ask you which force drove the boat, you might say wind. However, wind is created when there is a difference between the temperature of two places. Hot air rises and expands, and cold air rushes to fill up the void. At the heart of this process is the Sun, which is the engine of all weather and nature on Earth. So, it is correct to say that the sails were solar-powered, even if indirectly.

- The sail allowed people to cross endless oceans, but that is pale in comparison to the vast space around us. The oceans are vast, but space is infinite in comparison. Nothing moves or changes there without purpose.

Our rockets are currently too slow and inefficient. Apollo and SpaceX missions, and almost all of the rest, rely on chemical fuel. They burn through massive amounts of propellant just to escape Earth's atmosphere. Even reaching our nearest planets remains a challenge with these.

- Scientists have discovered new ways, like nuclear propulsion and ion rockets, but each one of them presents massive challenges. Proxima Centauri, the closest star system, is 4 light years away, but with today's rockets, that might as well be infinite.

Ion drives are really efficient and accelerate continuously, but there is a major issue with them. They rely on xenon gas, which is very limited, and we simply can't carry enough to reach another star.

- The space probe was the fastest craft ever produced by us, travelling at 690,000 kilometers per hour towards the Sun. In terms of outer space, Voyager 1 keeps the record of a speed of 61,000 km/hr. At that speed, Proxima Centauri will take 17,000 years to reach!

However, the answer isn't larger rockets or engines; it's the sailboat. We have even started working on it. The technology is called a solar sail.

- This isn't pushed by engines, but by the gentle push of light. The Sun emits photons (tiny particles which do not have mass, yet they have energy and momentum). They still exert force when they collide with any surface.

When we build a highly reflective solar sail, the photons transfer their momentum to the craft. Unlike chemical rockets which use fuel, they use the constant photons emitted by sunlight. The thrust is really slow in the beginning, but it accelerates as the momentum continues.

- They must be ultra-light, ultra-strong, and highly reflective, because absorption won't generate thrust and would heat up the sail instead of pushing it ahead. Direction works by moving the sail's angle, so the sail acts as the rudder, the fuel, and the compass.

This isn't science fiction. JAXA deployed IKAROS in May of 2010. IKAROS reached speeds of thousands of kilometers per hour while emitting data, and eventually flew beside Venus and continued transmitting data for 5 years. With a gravity assist from Venus, the craft reached speeds of 12,000 km/hr.

- In the year 2019, The Planetary Society used a solar sail in Earth's orbit and even carried out precise maneuvers without any engines. In September of 2024, NASA deployed ACS3, successfully deploying its 9-meter-wide solar sail. The acceleration is slow, but the spacecraft continues to emit valuable data.

However, something is missing, even now. These sails still can't reach stars fast enough with the Sun alone. Therefore, we need lasers from Earth.

- Project Breakthrough Starshot, proposed in 2016 by Yuri Milner, Stephen Hawking, and Mark Zuckerberg, aims to push the sail to 20% of the speed of light — fast enough to reach Proxima Centauri in 20 years instead of thousands.

The problem is cost. Building the initial lasers would take billions. While we could just rely on the Sun, the lasers really give the push. We could use satellite-based lasers, but an Earth-based laser array would be unmatched, which can generate immense power.

- The idea is to magnify the effect of photons using lasers, and concentrate them on a small area, which would provide immense propulsion. Once the spacecraft reaches deep space, the Sun's acceleration becomes weak. Therefore, we need to accelerate the sail a lot faster before it leaves the solar system.

Once the sail reaches its target system, slowing down is thought to be relatively easy. Just turn the sail around, and the particles which once accelerated it could now act as a brake.

- We would eventually travel there ourselves, and to achieve that, the sail would have to be enormous, which could only be manufactured and assembled in space. The ship would need massive radiation shielding, and such a long period in low gravity will pose serious health risks.

These sails shall be our ticket to the stars.

-

THE ENGINE INDUSTRY

Nuclear Propulsion

One of the major obstacles separating us from sending probes far from Earth is speed. To develop faster rockets, NASA is going nuclear. This was meant to be a mission with DARPA (a U.S. government research agency). However, they soon exited from the mission due to heavy costs compared to benefits, and shared their knowledge and understanding developed with NASA and other defence agencies.

- The rocket introduced by NASA later on promised to help us reduce transit times to Mars to just 2 months, not an easy task either, but better. The technology has come from Howe Industries, and it is called the Pulse Plasma Rocket (PPR). This plans to provide shielded, fast travel to Mars, meaning that they are trying to avoid radiation during interplanetary travel.

Now, this technology is not easy to understand or use, especially the way it is presented and the data available. It is similar to designs from NASA back in the 1950s and 60s for Orion. The rocket blasts tiny nuclear pulses as it moves, which propel the ship forward. This may sound insane, but it works.

- It is called Pulsed Fission-Fusion (PUFF), and involves small nuclear reactions. Radioactive material, uranium or deuterium, is fired down the barrel using an electromagnetic launcher. As the pellets travel down the barrel, they are hit by a pulsed laser, which injects neutrons that eventually superheat the material. This triggers a nuclear fission reaction in uranium, which splits the atom.

The energy released then triggers a nuclear fusion reaction in deuterium and tritium, like a cascading release of energy that instantly generates a massive expansion of supercharged particles, known as plasma.

- At the end of the rocket, there is a magnetic nozzle which repels all the charged particles out of the back and pushes the rocket forward. The nozzle also captures energy from the reaction for the next pulse.

The point of all of this complex and dangerous technology is to develop systems with both speed and efficiency at the same time, which is not possible with a conventional engine. You could have a chemical engine like Raptor from SpaceX, but it will burn through thousands of tons of fuel in seconds, or an electric engine like a Hall-effect thruster, which can run for long durations with just a few tons of xenon gas but takes a lot of time to accelerate.

- The pulse thruster can provide 100,000 newtons of thrust with 5,000 seconds of specific impulse. The SpaceX Raptor has around 330 seconds of specific impulse, which is much lower.

We can afford to add more shielding and safety to the rocket to an extent which would be impossible for conventional ones. NASA has moved the PPR rocket idea to Phase 2 to assess it further and gather crucial knowledge.

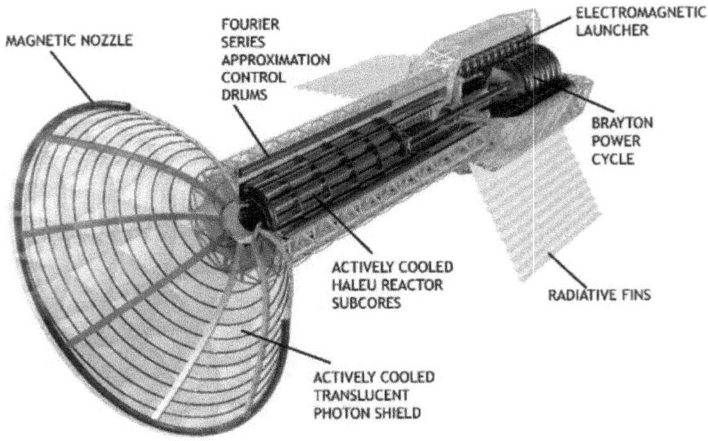

--By NASA

MANUFACTURING

IN SPACE

The next industrial revolution, or the one after AI, will be from space. Some things are built at a scale just not possible to manufacture on Earth, as Earth is densely packed. The microgravity and vacuum can provide a superiority to products never achieved before.

- This will impact pharmaceuticals first, as medicine is one of the most important fields of humanity. Many diseases could hopefully be cured using this technology, and treatments could become much faster and better.

"Made in Space" objects might appear, and they will be an instant success for the first years because of general public interest and curiosity. It might take a decade or so to normalise space products and scale them up, so the products will be rare and collection pieces will most probably circulate.

- Earth is packed with so many things, and in each breath you take, you inhale harmful particles or viruses. However, the vastness of space is empty, and therefore has little to no particles.

In delicate industries like semiconductors, where even a small particle of dust can cause a massive loss and make the whole batch useless, the vacuum of space offers a major advantage.

- In microgravity, protein crystals grow larger and more uniform because gravity-driven sedimentation is removed, facilitating better X-ray crystallography for drug development.

Advanced fiber optic cables (ZBLAN) can be manufactured in space, which can achieve lower signal loss than Earth-based silica cables, which have impurities and imperfections.

- Due to high surface tension and no gravity, metal components can form perfect spheres, impossible on Earth where gravity distorts shape during cooling.

We can also replicate parts of our body. The BioFabrication Facility on the ISS has been set up to test this. When we try to replicate human tissues on Earth, gravity causes soft, weak, and unique structures to collapse. In space, building organs could become a new reality.

- Overall, space manufacturing can be revolutionary, but of course comes with its own challenges. How does one stop small components like semiconductors from floating away and still produce them in mass?

However, we can still accomplish it and establish ourselves in many industries, and potentially save many lives, for which all of this is worth it.

3D bioprinter

Overall, space is heading towards a really positive outlook. You can really see how it can impact our day-to-day lives, and I'm sure it will. Space isn't just about gaining something, but about exploring the endless universe and assessing human capabilities, working towards something we should.

Conclusion

In the end, it all will make sense. Though we might believe that it's not possible, permanent human settlement in space will be a reality in a few years, decades, or even a century. It might even happen within your lifespan, and we could be the ones who are the settlers.

- The health risks that evolve with other planets will change every time, but one thing remains certain: our resilience and the wish to never give up.

The Europeans crossed the world with boats and explored places from which there was no certainty, and agriculture was discovered. We wandered off to the Moon and now to Mars. There will always be obstacles, and changes in muscles and the way our bodies function are necessary.

- Change is survival, and those who don't grow finish themselves, as the dinosaurs exhibit.

The Earth is made of space, and there must be more of the things that made up Earth scattered around in the universe, just unseen to us.

- The main risks that space poses are related to gravity itself, and our bodies will adapt in a generation or so.

A century or two from today, we will have bases on Mars (Oh! Sorry, metropolitans) and maybe Jupiter or Saturn too.

- The resources that are limiting us on Earth are basically unlimited in the universe, and we need to explore to gain access to them. The problem isn't our population; it's our reach.

Some issues like homelessness, resource limitation, and others can be easily resolved using space mining.

Bibliography

This is the list of references kept in mind and used while writing this book-

1. NASA[https://www.nasa.gov/budgets-plans-and-reports/]
2. ISRO[https://www.indiabudget.gov.in/doc/eb/sbe95.pdf]
3. From Earth to Mars by Himank Mittal
4. Tsiolkovsky, K. E. (1903). The Exploration of Cosmic Space by Means of Reaction Devices. [Source for the Rocket Equation]
5. Sutton, G. P., & Biblarz, O. (2016). Rocket Propulsion Elements. [Link: Wiley Technical Library]
6. NASA Science. Basics of Space Flight. [Link: NASA.gov]
7. Bate, R. R., Mueller, D. D., & White, J. E. (1971). Fundamentals of Astrodynamics. [Link: Dover Publications]
8. Anderson, J. D. (2010). Fundamentals of Aerodynamics. [Link: McGraw Hill]
9. Young, W. C., & Budynas, R. G. Roark's Formulas for Stress and Strain. [Link: AccessEngineering]
10. Curtis, H. (2013). Orbital Mechanics for Engineering Students. [Link: Elsevier]
11. National Academies Press. Radiation and the International Space Station. [Link: NAP.edu]
12. European Space Agency (ESA). The Oberth Effect: Boosting Efficiency. [Link: ESA.int]
13. SpaceX Technical Documents. Falcon 9 User's Guide (Max Q and Re-entry). [Link: SpaceX.com]
14. Halliday, D., Resnick, R., & Walker, J. Fundamentals of Physics (Thermodynamics and Momentum). [Link: Wiley]
15. The Feynman Lectures on Physics. Vol 1, Ch 10: Conservation of Momentum. [Link: FeynmanLectures.caltech.edu]
16. United Launch Alliance. Atlas V Mars 2020: Mission Overview and Launch Vehicle Specifications. Centennial, CO: ULA Technical Press, 2020. www.ulalaunch.com
17. Farley, K. A., and K. H. Williford. "The Mars 2020 Mission: Science Objectives and Jezero Crater Landing Site." Space Science Reviews 216, no. 142 (October 2020). link.springer.com
18. National Aeronautics and Space Administration (NASA). Mars 2020

Perseverance Landing Press Kit. Pasadena, CA: Jet Propulsion Laboratory, 2021. www.jpl.nasa.gov

19. Schoenenberger, Mark, et al. MEDLI2 Reconstruction: Aerothermodynamics and Trajectory of the Mars 2020 Entry. NASA Technical Report NTRS-20210024320. Hampton, VA: Langley Research Center, 2021. ntrs.nasa.gov
20. Grip, Håvard F., et al. "Flight Dynamics and Control of the Ingenuity Mars Helicopter." NASA/JPL Technical Archives, April 2021. mars.nasa.gov
21. Iannucci, S. A., and M. C. Pajereck. "Mars 2020 Entry, Descent, and Landing System Software Implementation." AIAA SciTech Forum, 2022. ntrs.nasa.gov
22. NASA Jet Propulsion Laboratory. "Terrain-Relative Navigation: Autonomous Landing Technology for Mars 2020." JPL Technology Highlights, February 2021. mars.nasa.gov
23. Francis, Raymond, et al. "Autonomous Robotics is Driving Perseverance Rover's Progress." Science Robotics 8, no. 79 (June 2023). www.science.org
24. Von Ehrenfried, Manfred. Perseverance and the Mars 2020 Mission: Follow the Science to Jezero Crater. New York: Springer Praxis Books, 2022. www.springer.com
25. NASA Space Communications and Navigation. The Deep Space Network: Basics of Interplanetary Communication. Washington, DC: NASA Headquarters, 2022. www.nasa.gov
26. National Aeronautics and Space Administration (NASA). Apollo 17 Press Kit. Release no. 72-220K. Washington, DC: November 26, 1972. www.history.nasa.gov [Source for Launch Trajectory and mission timeline].
27. Lyndon B. Johnson Space Center. Apollo 17 Mission Report. JSC-07904. Houston, TX: National Aeronautics and Space Administration, 1973. ntrs.nasa.gov [The definitive technical record of the Command Module and Lunar Module performance].
28. Jones, Eric M., ed. The Apollo 17 Lunar Surface Journal. NASA History Division, 1995. history.nasa.gov [Annotated transcripts of all technical activities at the Taurus-Littrow landing site].
29. NASA Manned Spacecraft Center. Apollo 17: Preliminary Science Report. NASA SP-330. Washington, DC: Scientific and Technical Information Office, 1973. www.lpi.usra.edu

30. Cernan, Eugene, and Donald A. Davis. The Last Man on the Moon: Astronaut Eugene Cernan and America's Race in Space. New York: St. Martin's Press, 1999. [Technical and personal account of the Lunar Module Challenger landing].
31. NASA Marshall Space Flight Center. Saturn V Launch Vehicle Flight Evaluation Report - AS-512 (Apollo 17). MPR-SAT-FE-73-1. Huntsville, AL: 1973. ntrs.nasa.gov [Source for the physics of the Final Saturn V night launch].
32. Wilhelms, Don E. To a Rocky Moon: A Geologist's History of Lunar Exploration. Tucson: University of Arizona Press, 1993. [Reference for the Landing Site Selection physics and geology].
33. Woods, David, and Ben Feist. The Apollo 17 Flight Journal. NASA History Division, 2020. history.nasa.gov [Step-by-step technical guide to the Trans-Lunar Injection and journey path].
34. Orloff, Richard W. Apollo by the Numbers: A Statistical Reference. NASA SP-4029. Washington, DC: NASA History Division, 2000. history.nasa.gov [Statistical breakdown of fuel consumption and mission weights].
35. Schmitt, Harrison H. "The Real World of the Apollo 17 Astronauts." In Lunar Sourcebook: A User's Guide to the Moon, edited by G. H. Heiken and B. M. French. Cambridge: Cambridge University Press, 1991.
36. National Aeronautics and Space Administration (NASA). Hubble Space Telescope: Service Mission 1 Media Reference Guide. Greenbelt, MD: Goddard Space Flight Center, 1993. www.nasa.gov [The primary technical source for the COSTAR optics repair that fixed the spherical aberration].
37. Villard, Ray, and Zoltan Levay. Hubble: 25 Years of Discovery. New York: Abrams, 2015. [A comprehensive guide to the Top Discoveries, including the expansion of the universe and exoplanet atmospheres].
38. Smith, Robert W. The Space Telescope: A Study of NASA, Science, Technology, and Politics. Cambridge: Cambridge University Press, 1989. [The definitive history of the Physics of the Mirror and the engineering challenges of a telescope in vacuum].
39. Space Telescope Science Institute (STScI). The Hubble Deep Field: Final Science Report. Baltimore, MD: STScI Press, 1996. www.stsci.edu [Technical record of how Hubble looked back 13 billion years in time].
40. Zimmerman, Robert. The Universe in a Mirror: The Saga of the Hubble Space Telescope and the Visionaries Who Built It. Princeton: Princeton

University Press, 2010. [Excellent for the spectroscopy and imaging physics used to find Dark Energy].

41. National Aeronautics and Space Administration (NASA). Cassini-Huygens: The Grand Finale Press Kit. Pasadena, CA: Jet Propulsion Laboratory, 2017. www.jpl.nasa.gov [The primary technical record for the Grand Finale Orbits and the physics of the spacecraft's atmospheric entry into Saturn].
42. Lebreton, Jean-Pierre, et al. "An Overview of the Descent and Landing of the Huygens Probe on Titan." Nature 438, no. 7069 (December 2005): 758–764. www.nature.com [The definitive scientific source for the Aeroshell Physics and the landing data from the surface of the moon Titan].
43. Grotzinger, John P., et al. "Mars Science Laboratory Mission and Science Investigation." Space Science Reviews 170, no. 1-4 (July 2012): 5–56. link.springer.com [The definitive technical and scientific overview of the Rover's Design, the Sky Crane landing physics, and the search for habitable environments in Gale Crater].
44. Dickson, Paul. Sputnik: The Shock of the Century. New York: Walker & Company, 2001. www.nebraskapress.unl.edu [A comprehensive history detailing the Soviet Union's development of the first artificial satellite and the immediate political and social upheaval it caused in the United States].
45. National Aeronautics and Space Administration (NASA). Artemis I Press Kit. Washington, DC: NASA Headquarters, 2022. www.nasa.gov [The primary technical overview for the inaugural flight of the Space Launch System rocket and Orion spacecraft, setting the foundation for long-term human exploration of the Moon].
46. Feinzeig, Sara Dina. The Impact of Space Travel on the Human Body. Touro Scholar, 2014. touroscholar.touro.edu [An analysis of the deleterious effects of microgravity on bone density, muscle atrophy, and cardiovascular fluid redistribution].
47. "NASA's 3D-printed Rotating Detonation Rocket Engine Test a Success." NASA Marshall Space Flight Center, 2023. www.nasa.gov/centers-and-facilities/marshall/nasas-3d-printed-rotating-detonation-rocket-engine-test-a-success/ [Details the testing and performance of a 3D-printed Rotating Detonation Rocket Engine (RDRE) for future lunar and Mars missions].

www.ingramcontent.com/pod-product-compliance
Lightning Source LLC
LaVergne TN
LVHW050936210926
848482LV00013B/637

* 9 7 9 8 9 0 3 9 2 8 0 2 6 *